Consensus and Controversies in Animal Biotechnology

Consensus & Controversies in Animal Biotechnology

An Interactive Legislative Approach to Animal Biotechnology in Denmark, Switzerland, and the Netherlands

Lonneke Poort

eleven
international publishing

Published, sold and distributed by Eleven International Publishing
P.O. Box 85576
2508 CG The Hague
The Netherlands
Tel.: +31 70 33 070 33
Fax: +31 70 33 070 30
e-mail: sales@budh.nl
www.elevenpub.com

Sold and distributed in USA and Canada
International Specialized Book Services
920 NE 58th Avenue, Suite 300
Portland, OR 97213-3786, USA
Tel: 1-800-944-6190 (toll-free)
Fax: +1 503 280-8832
orders@isbs.com
www.isbs.com

Eleven International Publishing is an imprint of Boom uitgevers Den Haag.

ISBN 978-94-90947-94-1
ISBN 978-94-6094-682-0 (E-book)

Printed in The Netherlands

Table of Contents

Part I

1	**Introduction**	**3**
1.1	Research Topic	3
1.2	Animal Biotechnology	7
1.2.1	Animal Biotechnology	7
1.2.2	A Complex Policy Problem with a Moral Impact	8
1.3	The Interactive Legislative Approach	10
1.4	Approach and Methods of Research	13
1.4.1	Definition of the Problem	13
1.4.2	The Countries	13
1.4.3	Outline	14
1.4.4	Methodology	15
2	**A Comprehensive Framework on the Quality of Law**	**17**
2.1	Introduction	17
2.2	The Framework	17
2.3	The Dimension of Fit	19
2.3.1	Fit with Law	19
2.3.2	Fit with Context	20
2.3.3	Coping with Problems	21
2.4	The Dimension of Political Morality	21
2.4.1	Substantive Justification	22
2.4.2	Popular Legitimacy	23
2.5	The Dimension of Legal Validity	24
2.6	The Framework	25
2.7	Concluding Remarks	26
3	**The Interactive Legislative Approach**	**27**
3.1	Introduction	27
3.2	Theoretical Backgrounds	29
3.3	The Various Interactive Approaches	31
3.4	The Characteristics of the Interactive Legislative Approach	37
3.5	Final Remarks	40

Part II

4	**Introduction to the Case Studies**	**43**
4.1	Introduction	43
4.2	Characteristics of a Legislative Approach	44
4.3	Conditions for an Interactive Legislative Approach	46
4.4	The Layout	48
5	**The Regulation of Animal Biotechnology in Denmark**	**49**
5.1	Introduction	49
5.2	The Regulation of Animal Biotechnology	49
5.2.1	The Legislative Procedure	50
5.2.2	The Legislation in Practice	54
5.3	Characteristic Elements of a Legislative Approach	55
5.3.1	Public Involvement in Legislation	55
5.3.2	The Role of Moral Values in Legislation	56
5.3.3	Legitimacy of Legal Norms	59
5.4	Danish Political Culture	61
5.4.1	Democratic Traditions	61
5.4.2	The Role of the Public	63
5.4.3	The Legislative Culture	70
5.5	Conclusion	73
6	**The Regulation of Animal Biotechnology in Switzerland**	**75**
6.1	Introduction	75
6.2	The Regulation of Animal Biotechnology	75
6.2.1	The Legislative Procedure	76
6.2.2	The Legislation in Practice	80
6.3	Characteristic Elements of a Legislative Approach	81
6.3.1	Public Involvement in Legislation	81
6.3.2	The Role of Moral Values in Legislation	83
6.3.3	Legitimacy of Legal Norms	85
6.4	Swiss Political Culture	87
6.4.1	Democracy and Federalism	87
6.4.2	The Role of the Public	90
6.4.3	The Legislative Culture	93
6.5	Conclusion	95

7	**The Regulation of Animal Biotechnology in the Netherlands**	**97**
7.1	Introduction	97
7.2	The Regulation of Animal Biotechnology	98
7.2.1	The Legislative Procedure	98
7.2.2	The Legislation in Practice	102
7.3	Characteristic Elements of a Legislative Approach	105
7.3.1	Public Involvement in Legislation	105
7.3.2	The Role of Moral Values in Legislation	107
7.3.3	Legitimacy of Legal Norms	109
7.4	Dutch Political Culture	110
7.4.1	Democratic Traditions	111
7.4.2	The Role of the Public	113
7.4.3	The Legislative Culture	115
7.5	Conclusion	117
8	**Reflections on the Case Studies**	**119**
8.1	Introduction	119
8.2	Characteristic Elements of the Interactive Legislative Approach	119
8.2.1	Co-Operative Effort	119
8.2.2	Use of Open Norms	122
8.2.3	Moral Values	123
8.2.4	Consensus	125
8.2.5	Legitimacy	127
8.2.6	Long-Run Problem Solving	129
8.3	Conditions	131
8.3.1	Democratic Traditions	131
8.3.2	Legal Culture	132
8.3.3	Ethics Committees	134
8.4	Difficulties for Dynamics	135
Part III		
9	**Consensus Reconsidered**	**139**
9.1	Introduction	139
9.2	Preliminary Remarks	140
9.2.1	The Context of Reconsideration	140
9.2.2	Defining Consensus	141
9.3	An Ethos of Consensus	146
9.4	An Ethos of Consensus Reconsidered	149
9.4.1	An Alternative?	149

10	**An Ethos of Controversies in a Two-Track Approach**	**151**
10.1	Introduction	151
10.2	An Ethos of Controversies	153
10.2.1	The Circumstances of an Ethos of Controversies	153
10.2.2	An Ethos of Controversies	156
10.3	The Two-Track Approach	158
10.3.1	Why is a Two-Track Approach Required?	158
10.3.2	The Legal Discourse vs. the Moral Discourse	160
10.4	A Track for Legal Norm Development	162
10.5	A Track for Moral Norm Development	164
10.6	Remaining Remarks	165
10.6.1	Critical Remarks	165
10.6.2	Two New Hypotheses	166
Bibliography		**167**
Parliamentary Documents		**177**
Appendix I		**179**
Index		**181**

Part I

1 Introduction

1.1 Research Topic

Advances in animal biotechnology have led to major growth in the human ability to control the process of life. Biotechnology is controversial. While it has led to various new developments in medical science that have had a positive influence on people's lives, it has also incited disagreement on various levels. With developments in biotechnology have come questions about the moral impact of new technologies. Biotechnology has brought to the forefront questions about how we, as human beings, should behave towards nature, towards animals, and towards each other.

In particular, developments in genetic modification and cloning have brought with them new moral challenges, and touch on existential values. The use of animals for biotechnology research and experimentation has disrupted traditional ideas about the moral status of animals. This issue was thrust into public consciousness when scientists at the Roslin Institute in Scotland cloned the sheep "Dolly" in February 1997. For the wider international public, this sheep represented a major breakthrough in scientists' attempts to interfere with the process of life. Although genetic engineering has a much longer history, this single sheep generated huge amounts of debate on the moral limits to interfering with life. This case gave rise to a worldwide call for the regulation of genetic engineering. However, the regulatory process was complicated by the potential moral implications of these developments, particularly with respect to the impact of gene-technology and the status of animals.

Both in academic studies and in legislative practice, there has been a growing interest in alternative legislative approaches for addressing morally controversial issues. In particular, communicative, symbolic or interactive techniques seem to hold promise as alternative ways of developing legislation on issues characterized by strong moral disagreements, such as life-altering technologies. There is not one communicative, symbolic or interactive approach, but rather various lines of research.[1] These lines have in common an interest in the role of communication processes in the development and implementation of

1 See for example on interactive, communicative and symbolic legislation N. Zeegers, W. Witteveen, and B. van Klink (eds.), *Social and Symbolic Effects of Legislation Under the Rule of Law,* Lewiston New York, The Edwin Mellen Press, 2005; W. van der Burg and F.W.A. Brom, 'Legislation on Ethical Issues: Towards an Interactive Paradigm', in *Ethical Theory and Moral Practice,* 3/1, 2000, pp. 57-75; J. Brunnee and S.J. Toope, *Legitimacy and Legality in International Law: An International Account,* Cambridge, Cambridge University Press, 2010, on responsive law P. Nonet and P. Selznick, *Law and Society in Transition: Toward Responsive Law,* New York, Harper Row, 1978; J. Black, 'Regulation as Facilitation', *The Modern Law Review Limited,* 1998.

legal norms, as well as an interest in the moral dimensions of law.[2] But they also exhibit significant differences. Some of the approaches are grounded in legal theory, others in disciplines such as sociology, political science, or ethics. A single ideal-typical model of the interactive legislative approach has therefore not yet been developed.

This book will examine three lines of research on the interactive legislative approach to come to an ideal-typical model of the approach that combines a descriptive perspective with normative claims about how best to engage with complex regulatory issues with a moral impact. The research refers throughout to *moral* issues and *moral* impact rather than *ethical* issues or *ethical* impact. Debates about the differences between moral and ethical issues or moral and ethical impact and the nuances they imply are beyond the scope of this research. Here, it is only necessary to highlight the complexity of developing regulations in the field of animal biotechnology, which is caused by, among other things, its moral impact. Bovenkerk argues that the biotechnology debate is both ethical and moral, implicating both fundamental views concerning what 'would constitute a desirable society, about what are proper relationships within society and with nonhuman nature, and even about our own nature',[3] as well as people's existential values. However, this research will use the term 'moral' to refer to both fundamental views and existential values. 'Moral issues' here are therefore issues that are characterized by fundamental disagreements about existential values derived from different views about what constitutes a desirable society, and what the relationship between humans and nonhumans ought to be. 'Moral impact' in this research will refer to an issue's impact on existential moral norms and values.

One of the lines of research that will be examined is a version of the interactive legislative model first developed by Vorstenbosch and Ippel and further elaborated by Van der Burg and Brom.[4] This line of research highlighted the ways in which legal discourse was moving toward a more interactive paradigm.[5] It argued that the traditional instrumentalist approach to regulation fell short when legislatures were confronted with the moral issues, and instead looked to symbolic interactionism as offering possibilities for building a bridge between the complexity of these moral issues and the need for regulation. According to Vorstenbosch and Ippel, legislatures can adequately address moral dilemmas only by adapting to social norms and values. They therefore emphasize social interaction between government and citizens, arguing that it can lead to the development of a legal framework that fits these norms and values.

---

2 W. van der Burg, 'The Irony of a Symbolic Crusade: The Debate on Opening Up Civil Marriage for Same-Sex Couples', in N. Zeegers, W. Witteveen, and B. van Klink (eds.), *Social and Symbolic Effects of Legislation Under the Rule of Law,* Lewiston, New York, The Edwin Mellen Press, 2005, p. 246.

3 B. Bovenkerk, *The Biotechnology Debate. Democracy in the Face of Intractable Disagreement,* Deventer, Springer, 2011, p. 2.

4 J. Vorstenbosch and P. Ippel, 'De weg naar de wet. Wetgeving over ethische kwesties: Een interactief proces'. As point of departure they use the ideas of Witteveen, Van Seters, and Van Roermund about symbolic legislation; W.J. Witteveen, P. Van Seters and G. van Roermund (eds.), *Wat maakt de wet symbolisch?*, Zwolle, W.E.J. Tjeenk Willink, 1991.

5 Vorstenbosch and Ippel, 'De weg naar de wet. Wetgeving over ethische kwesties: Een interactief proces', pp. 53-54.

Van der Burg and Brom developed this argument more thoroughly. They expand on Vorstenbosch and Ippel's work by presenting some claims that they argue can function as starting points for both descriptive and normative analysis of legislation on moral issues.[6] Van der Burg and Brom begin their argument for an interactive model with the claim that where clear moral norms do not exist, the traditional legal paradigm no longer holds. In this case, they argue, it is necessary to move toward a more interactive paradigm. Other theorists follow Van der Burg and Brom's ideas. Grotefeld, for example, applies their ideas to the regulation of embryo selection in Switzerland. In doing so, he presents similar claims regarding the possibility that an interactive process could facilitate ongoing debate and reflection on moral issues. However, Grotefeld further argues for an understanding of political legitimacy that views the principle of consent not on neutral grounds as presented in the liberal tradition, but rather as also able to reflect non-neutral arguments. Neutrality, here, is not the core precondition for guaranteeing the recognition of citizens as free and equal. Instead, citizen involvement and the potential for reflecting on non-neutral arguments ensure that citizens can act freely and equally.[7]

The research presented in this book will take Van der Burg and Brom's ideas as a starting point for descriptive analysis. Their description of a situation in which the traditional legal model does not hold applies aptly to the case of animal biotechnology: clear moral norms about human obligations to animals cannot be identified in current social reality, which poses difficulties for traditional legal paradigms.[8] However, as their model is limited and not further developed into a clear ideal-typical model, it requires further elaboration. This book will therefore broaden the theoretical perspective by exploring other lines of research.

The second line of research this analysis of interactive approaches to law will explore is exemplified by the work of Julia Black. Black argues for a concept of regulation as facilitation.[9] In analysing the regulation of gene technology issues in the UK, Black distinguishes three views of law: the technocratic view, the modified technocratic view, and view of regulation as facilitation.[10] In general, technology issues are addressed via a technocratic approach in which scientists are dominant in decision-making.[11] She argues for a different

6 Van der Burg and Brom, 'Legislation on Ethical Issues'; Van der Burg, 'The Irony of a Symbolic Crusade'.

7 S. Grotefeld, 'Self-Restraint and the Principle of Consent', in *Ethical Theory and Moral Practice*, 2000, pp. 77-92; S. Grotefeld, 'Wie wird Moral ins Recht gesetzt?', in *Archiv. für Rechts- und Sozialphilosophie*, 2003, pp. 299-317.

8 See also below, Section 1.3.

9 Black, 'Regulation as Facilitation'.

10 *Ibid.*, pp. 652-655.

11 Despite the fact that the UK has a different legal system, and it may thus cause difficulties to make a comparison of regulation, Blacks conception of regulation as facilitation offers a substantive contribution to developing the model of the interactive legislative approach. Both theoretical explorations share interest in the communicative processes of law and mechanisms to stimulate interaction. Their theoretical background is thus comparable, and therefore Blacks ideas can contribute to developing an ideal-typical model of the interactive legislative approach.

view of law, which she terms regulation as facilitation. Technology issues should not be governed only by experts' opinions, but also by the concerns and opinions of lay people. Because of this, Black argues, legislators need to operate in such way that all viewpoints are included in the decision-making process. The function of law is not to control the risks of technology, but to facilitate the integration of all viewpoints concerning this technology. Black also emphasizes additional functions of law, such as the cognitive and communicative functions.

Third, this research will examine Brunnee and Toopes exploration of the interactional international theory of law.[12] In their international theory, Brunnee and Toopes emphasize the need for interaction to sustain the legality of norms. This pragmatic theory of how international law is made is inspired by Fuller's interactional theory.

By comparing these three lines of research, this book will reconstruct an ideal-typical model of the interactive legislative approach that takes into account the demands of the interaction, regulation as facilitation, and the Fullerian model. This ideal-typical model will consist of a set of theses, hypotheses, and assumptions regarding the characteristics and incentives of the interactive legislative approach.

In addition to developing an ideal-typical model of the interactive legislative approach, this book aims to analyze its usefulness for dealing with complex issues with strong moral impacts, such as animal biotechnology. Part I will therefore begin by developing a framework on the quality of law. This framework includes various dimensions on which the quality of law may depend. The framework will function as a baseline from which subsequent chapters will evaluate the ideal-typical model of the interactive legislative approach.

Furthermore, this books aims to demonstrate the practical value of the model it develops. As a result, Part II of this analysis will examine the regulation of animal biotechnology in three countries: Switzerland, Denmark, and the Netherlands. The outcomes of these case studies pose several challenges to the ideal-typical interactive legislative model. Part III explores these challenges, and criticizes some significant characteristics of the interactive legislative approach. In particular, it questions the model's assumptions that consensus should function as an aim for structuring ongoing norm development. It therefore turns to an application of these criticisms to refine and reconstruct the ideal-typical model. This new model abandons the focus on consensus and gives conflict a more prominent role.

The remainder of Chapter 1 will explore the topic of this research more thoroughly. It will explain the applications of the technology, as well as animal biotechnology as a policy problem that requires regulation. Additionally, this chapter will present an argument about why traditional instrumental legislative approaches no longer hold, and why the interactive legislative approach may offer an adequate alternative. The chapter will then conclude by outlining the approach of this research, including its method and structure.

12 Brunnee and Toope, *Legitimacy and Legality in International Law.*

1.2 Animal biotechnology

1.2.1 Animal Biotechnology

In answering the question 'what is animal biotechnology?', this section starts from Brom's analysis of animal biotechnology as a moral problem.[13] In his dissertation, Brom explains and defines animal biotechnology from the perspective of a layperson:

Animal biotechnology is the systematic application of knowledge on 'new biology' to animals to purposively direct their biological processes.[14]

This knowledge on 'new biology' in general refers to gene technology. Gene technology applies knowledge of the structure of genetics to influence and direct biological processes. A more detailed description is given by the Cartagena Protocol on Biosafety, which defines biotechnology as: 'the application of: a) *in vitro* nucleic acid techniques, including recombinant deoxyribonucleic acid (DNA) and direct injection of nucleic acid into cells or organelles, or b) fusion of cells beyond the taxonomic family, that overcome natural physiological reproductive or recombination barriers and that are not techniques used in traditional breeding and selection.'[15] A full explanation of the complexity of gene technology is not necessary here. However, this section will give some examples of the application of the technology in order to give the reader a better idea of their moral impact and the complications associated with regulating these issues.

A first example of gene technology is the use of so-called knock-out mice. Genes are responsible for the production of proteins. Proteins influence the functioning of the body. To understand which gene is responsible for the production of a specific kind of protein and thus indirectly for a specific function, scientists perform tests with knock-out or transgenic mice. Knock-out mice are created by implanting cells in which a certain gene is 'knocked out' or changed into mice embryos. Researchers then examine the mice to see what the function of that gene is.[16] With these biotechnological procedures the researchers gain insights into hereditary diseases. For example, knock-out mice are commonly used for biomedical research about cancer. This use of gene technology has positive benefits for human beings, as these tests can contribute to the development of better treatments for these diseases. However, these results require the instrumental use of animals for the benefit of human beings. The mice themselves often become ill with cancer or hereditary

13 F.W.A. Brom, *Onherstelbaar verbeterd. Biotechnologie bij dieren als een moreel probleem,* dissertation, Assen, Van Gorcum, 1997, Ch. 2.

14 *Ibid.*, p. 46 (translated into English by the author).

15 Secretariat of the Convention on Biological Diversity, 'Cartagena Protocol on Biosafety, Article 3 (I)', 2000; Bovenkerk, *The Biotechnology Debate.*, p. 21.

16 The procedures are much more complex as I explain here, the cells need to be activated and the production of proteins is visualized by injecting fluorescing material. See for more information <www.kennislink.nl> and COGEM, *Trendanalyse Biotechnology*, 2004.

diseases and eventually die. On the other hand, because the use of gene technology makes these mice more suitable for tests, fewer animals are required overall.

A second example involves xeno-transplantation.[17] Xeno-transplantation involves the transplantation of organs from one species to another. This technology is relatively new and has inspired wide resistance. This technique has great potential benefits for extending and improving human life. Insufficient numbers of organ donors means that many individuals who could have been treated die due to lack of supply. Xeno-transplantation may offer a solution. For example, pig organs are often suitable for human transplantation as they are similar in size. However, there is a high risk of rejection because human and pig immune systems differ. In order to decrease these risks, the genes of potential donor pigs are modified to produce more human-friendly proteins. This makes the organs of the modified pigs more suitable for transplantation into human bodies as the chance of rejection decreases.

Despite these potential benefits, the technology is still controversial. Xeno-transplantation poses various health risks for human beings, as donor pigs may be carriers of diseases with unknown effects for human beings. Furthermore, many commentators strongly resist the idea of transplanting animal tissues into human beings. This type of transplantation raises strong questions about the moral positions of both humans and animals and the relationship between them.

1.2.2 A Complex Policy Problem with a Moral Impact

Animal biotechnology is a complex policy problem characterized by a strong moral impact, uncertainty, and many different perspectives.[18] The current social framework of norms and values falls short in dealing with new technological developments that complicate the status of animals. At the same time, new norms and values have not yet crystallized. In the case-study countries (Switzerland, Denmark, and the Netherlands), the complexity of this policy problem is influenced by rapidly changing technological possibilities, concerns about the social and moral impact of using these technologies, and fundamental disagreements about values, scientific methods, empirical facts, and definitions.[19]

As mentioned in the previous section, the potential applications of technologies such as genetic modification and animal cloning could increase human quality of life in several ways. The use of modified animals can improve medical research and contribute to major breakthroughs in medical science such as developing cures for life-threatening diseases. Nevertheless, the uncertainties and concerns surrounding the use of these technologies make their introduction a controversial issue.[20] According to Bovenkerk, animal

17 <www.allesoverdna.nl>; COGEM, *Trendanalyse*, 2004.

18 J.D. Moreno, 'Biotech is thrusting us into new political territory', *newscientist.com Issue 2879*, 28 August 2012.

19 On the moral impact of animal biotechnology in the Netherlands, see Brom, *Onherstelbaar verbeterd*.

20 B. Bovenkerk and L.M. Poort, 'The Role of Ethics Committees in Public Debate', in *International Journal of Applied Philosophy*, 22/1, 2008, pp. 19-36; Moreno, 'Biotech is thrusting us into new political territory'.

biotechnology is a field in which intractable moral disagreements exist.[21] She refers here to various disagreements that touch on fundamental values and worldviews as well as certain biases. Referring to a framework designed by Hisschemoller and Hoppe for defining policy problems, Bovenkerk defines animal biotechnology as an 'unstructured problem'.[22] A problem is unstructured if it lacks both factual and normative consensus. As a consequence, unstructured problems are complex and difficult to regulate.

The complex structure of animal biotechnology is characterized by three main elements: uncertainties about knowledge, rapid changes, and a profound moral pluralism. First of all, there is a lack of knowledge about the possibilities, the potential impacts, and the consequences of animal biotechnology both at this moment and in the near future. It is possible to anticipate certain scenarios concerning the impacts and consequences, but there may also be various unforeseen facts that could not be taken into account beforehand. This lack of knowledge causes uncertainty about the applications of (and thus the justifications for) animal biotechnology. Clear-cut regulation in the short term will therefore always be based on incomplete information.

Second, animal biotechnology is characterized by the rapid development of new applications and possibilities. Even if we could fully understand the policy problem and bridge the lack of knowledge, the rapid pace of developments would bring new uncertainties in the near future. New possibilities involve new fears, changing risks, and a change of social context. Furthermore, we should question whether we want certain applications and possibilities. It would be unrealistic to anticipate each possible scenario in policy- and decision-making in the legal context or to clarify our judgements about these scenarios before they occur. Clear-cut regulation cannot keep up with technological changes and moreover cannot justify them. A legal framework that consists of clear concrete norms leaves no room to respond to changes or be flexible towards new applications.

A third characteristic issue of animal biotechnology is a pervasive moral pluralism with regard to the status of animals and the range of applications of animal biotechnology. This moral pluralism works in complex ways. On the one hand, conflicting norms and values are rooted deeply in the beliefs and convictions of different groups in society, and therefore moral pluralism is unavoidable. In a nutshell, these conflicting norms and values involve the extent of animal integrity or animal dignity, the acceptability of the instrumental use of animals, the significance of the slippery slope-argument, concerns about the naturalness of organisms created using genetic engineering, and references to God as the creator of nature. On the other hand, these deeply rooted beliefs and convictions are constantly challenged by new developments in society and technology. In addition, the foundations of moral convictions regarding animal biotechnology are as yet unknown. And as far as there are to a certain extent foundational beliefs about the status of animals, there is no consensus about their application. All in all, these difficulties make animal

21 Bovenkerk, *The Biotechnology Debate,* Ch. 1.

22 M. Hisschemöller and R. Hoppe, 'Coping with Intractable Controversies: The Case for Problem Structuring in Policy Design and Analysis', in *The International Journal of Knowledge Transfer and Utilization,* 1995, pp. 40-60.

biotechnology a field characterized by intractable moral disagreements – a term that Black also uses to define issues of genetic engineering. The complexity brings in a language problem as well. Various disciplines are involved or concerned with the issue of animal biotechnology. Each discipline has its own cognitive framework. Bridging the differences in understanding makes the problem even more complex, and perhaps even intractable.[23]

As a result, regulating these issues is not easy. On the one hand, animal biotechnology is still developing, and there is much uncertainty about future possibilities. It is complicated to set concrete legal rules when it is not clear what to expect. Each new development brings new challenges to existing moral values. And the lack of consensus on the moral impact of animal biotechnology makes it difficult to establish legitimacy for clear-cut regulations and complicates the enforcement of the rules.

On the other hand, however, this complex policy problem needs coordination, regulation, and perhaps limitation before these conflicts come to a head and there is no way back from certain developments. A legislative strategy focused on defining clear-cut rules is poorly suited to dealing with the problematic issues of animal biotechnology. The characteristics of this issue require an alternative approach in which the various considerations can be weighed.

1.3 THE INTERACTIVE LEGISLATIVE APPROACH

As discussed in Section 1.1, this book will analyze three lines of research in order to develop an ideal-typical model of the interactive legislative approach: Van der Burg and Brom's interactive paradigm, Black's regulation as facilitation, and Brunnee and Toope's interactional international theory of law. Chapter 3 will present a thorough discussion of these lines of research. However, this section will provide a brief introduction to their use in the rest of this book.

Due to its focus on moral complexity, the ideal-typical model will use Van der Brug and Brom's model as its foundation. Van der Burg and Brom's interactive legislative approach attempts to address the difficulties that complex moral issues pose for political institutions and legal regulations. It focuses on the communicative dimension of legislation. This approach is built on the idea that interaction between the legislature and the public in a horizontal process of decision-making will result in more adequate norms. Interaction can improve both the legitimacy and sufficiency of legal norms. Moreover, the development of legal norms and norms 'on the ground' can reinforce each other when practitioners are involved in the legislative process. The use of open norms and the involvement of a broad set of actors after implementation can stimulate an ongoing process of norm development that continues as long as norms are not yet clear.

Van der Burg and Brom argue that the interactive approach is not only a response to new issues confronting the law, but also reveals that the character of law itself is changing.

23 Black, 'Regulation as Facilitation', p. 622.

They begin their argument by elaborating the pitfalls of the traditional paradigm, which is based on two basic assumptions that no longer hold for complex moral issues. First, the traditional paradigm assumes that there exist 'clear moral norms, belonging to either positive or to critical morality', and that 'law has such an authority or force that giving moral norms a legal status is an effective method to have citizens conform to those moral norms'.[24] These assumptions are no longer valid in modern society, as a result of three relevant changes.[25] To begin with, a process of horizontalization of power relations is emerging. This horizontalisation can be seen in society in general, as well as in the legal relationship between the state and private actors, and on the international level.[26] The authority of the law can no longer be taken for granted, since society is too complex to enforce legal norms. Moreover, legal norms that conflict with the moral or social norms of the people or society will face enforcement difficulties and may even remain a dead letter. The voluntary cooperation of citizens (or for international law, states) is, therefore, needed.

Second, there has been a change towards individualisation. This process can be seen in the number of different conceptualisations of the problem that is at stake, and about the solutions for this problem.[27] Additionally in the field of morality: there are no longer indisputable moral norms that can give guidance on every issue. Instead, moral issues are nowadays characterized by moral pluralism and rapid changes for which existing norms are inadequate.[28]

A third change concerns the dynamics of society. Black refers here to rapid developments in the field of technology. The law as well as morality have to deal with these new issues and the rapid changes that society faces. As a result, the law has become less static, and more dynamic than in previous generations. Brunnee and Toope argue that legal norms run the risk of losing commitment and becoming dead letters if they deviate from social practice. Interaction must continue to maintain the legal norm's influence. This is especially true in the context of international law, where legal obligation cannot be forced.

According to Van der Burg and Brom, these difficulties concerning the assumptions of the traditional paradigm require the development of new perspectives in both the legislative and the implementation process. The interactive approach, they argue, provides a better solution in the case of complex moral issues. First of all, moral issues impact the moral convictions and ideals of citizens. In a modern pluralist democratic society, citizens do not want to be forced by law to change their convictions and their ideals. Involving people in the development of the legal rules that concern their convictions and ideals makes it easier to accept those rules. Interaction will contribute to commitment of the rules.[29]

24 Van der Burg and Brom, 'Legislation on Ethical Issues', p. 59.
25 Van der Burg, 'The Irony of a Symbolic Crusade', pp. 248-249.
26 Brunnee and Toope, *Legitimacy and Legality in International Law*, pp. 5-9.
27 Black, 'Regulation as Facilitation', p. 621.
28 Van der Burg and Brom, 'Legislation on Ethical Issues', pp. 60-61.
29 Brunnee and Toope, *Legitimacy and Legality in International Law*, Ch. 1; L.L. Fuller, *The Morality of Law*, rev. edn., New Haven, Yale University Press, 1969, p. 221, p. 237.

Second, individuals' moral convictions are further concretized in practical situations. It is therefore preferable to leave room in the legal framework so that laws can be adapted to concrete situations. Third, interacting with the actors involved in a particular field allows the legislature to gain a more comprehensive understanding of reality and the moral norms that have already developed. Fourth, a legislature cannot capture all moral concerns and viewpoints, and should therefore consult experts who have insights into moral decision-making. Additionally, Grotefeld argues, a horizontal and inclusive decision-making process improves political legitimacy.[30]

Van den Burg and Brom define two main normative ideals of the interactive legislative approach: interaction and dynamics. Interaction is defined as decision making on the horizontal level. Dynamics should be understood in light of, referring to an ongoing process of norm development. Some scholars have criticized this approach. To begin with, interaction and ongoing processes may be inefficient. The interactive approach is characterized by broad participation in the legislative process and the use of open norms that may have an expressive and/or a communicative function, but leave room for further development. This process, which involves more actors than a traditional approach, will usually take longer. Second, including groups of involved actors in the legislative process can create difficulties with respect to democratic legitimacy, since decisions are not made solely by democratically elected representatives. Additionally, the use of open norms may undermine legal security and endanger norm conformity, since these norms are always open to further interpretation.

This book contends, however, that these objections are outweighed by the advantages of that a change of perspective could bring to the legislative process. These efficiency, legitimacy, and conformity problems can be tackled in three ways. First of all, the law will most probably be more effective and useful in the field if it is oriented to practice. Second, interaction ensures broader public support, since those who are working in the field are involved in norm development. And third, where moral norm development and legal norm development go hand in hand, actors working with legal rules in the field are likely to be in conformity with the new legal rules before these rules are enforced at all.[31]

From this perspective, not only the legislative process, but the implementation process, too, should be structured interactively. An interactive process can facilitate ongoing moral debate and reflection on the issues at stake. Moral consensus (which is lacking), then, is not presumed as a starting point, but is rather one of the aims of debate. Moral norms and values are developed through ongoing reflection and engagement during the implementation process, which can eventually lead to a consensus or to crystallization of the various differences that characterize the moral issues at stake.[32]

The interactive and dynamic character of this approach to legislation seems to address all three characteristic issues of: uncertainty, rapid change and moral pluralism. These characteristic issues require flexibility which the dynamic element of the interactive approach claims to offer. Furthermore, these characteristic issues require involvement of

---

30 Grotefeld, 'Self-Restraint and the Principle of Consent', pp. 77-92.
31 Van der Burg and Brom, 'Legislation on Ethical Issues', pp. 63-64.
32 *Ibid.*, pp. 64-67.

those actors to whom those norms address. Horizontal decision-making facilitates participation of those actors. This research will analyze the interactive approach as it applies to this context and examine its practical value.

1.4 Approach and Methods of Research

1.4.1 *Definition of the Problem*

The aim of this research is to develop an ideal-typical interactive model of law. As noted above, there is not one single communicative, symbolic or interactive approach to legislation, but instead different lines of research with different initial research interests. It will do so by exploring those various lines of research and reconstructing their claims in one ideal-typical model consisting of a set of descriptive theses, assumptions and hypotheses. Furthermore, the reconstruction will incorporate an examination of the model's practical value.

While starting from one prominent line of research introduced in the previous section, this book will broaden the model by taking on an international theoretical perspective and examining the model's practical value for addressing legal issues with a strong moral impact in three different countries. This international perspective will make it possible to discern how interactive legislative models are used in several countries. This will additionally contribute to identifying the specific conditions under which the interactive legislative approach can function adequately. Furthermore, this international perspective will provide a vantage point from which to criticize the approach and discern whether a further reconstruction of the theoretical model is required.

1.4.2 *The Countries*

This section will explain the choice of the three countries. In general, there were four criteria for the selection of the countries. First, the countries had to be modern democratic societies with various instruments of public participation. Without such instruments it would be difficult to establish a process of horizontal decision-making in which citizens and practitioners could participate. Second, the countries had to have existing or draft regulations concerning animal biotechnology. Third, the countries had to be heterogeneous societies with a plurality of identities, leading to a profound moral pluralism regarding complex issues with strong moral impacts. And fourth, the tenets of an interactive legislative approach had to be recognizable in the legal culture of the country, at least in academic debates.

In Switzerland, as in the Netherlands, animal biotechnology is regulated by a licensing procedure. In both countries, an ethics committee is appointed to advise the legislature on a case-by-case basis. This advice is public. In Switzerland, the regulation of animal biotechnology is the result of two referenda in which the Swiss people could vote on including protection for the 'dignity of living beings' (*Würde der Kreatur*) in the Swiss constitution. Here, we can recognize the Swiss tradition of direct democracy. The Swiss government also actively sought methods to stimulating public debate. In the Netherlands, the government also sought to stimulate public debate. The Dutch ethics committee had a prominent role in the legislative process as well as in public debate. In the Netherlands, licensing hearings are public and therefore incorporate a strong element of public participation. Both Switzerland and the Netherlands have regulations on animal biotechnology and can be characterized as modern democratic societies that have instruments for public participation.

Denmark is also an interesting case study because of Danish experience with public consultation. Denmark is known for its methods of stimulating public debate and public involvement, for example through consensus conferences. The legislation on genetic modification and cloning of animals mandated the assembly of a committee to decide on individual applications on a case-by-case basis. These decisions are (afterwards) open to the public. Furthermore, despite the prominent utilitarian flavour in Danish politics, animal integrity is explicitly referred to as one of the relevant criteria in the legislative process on the regulation of animal biotechnology.

The three countries also fulfil the other criteria for selection of the case studies. These countries are heterogeneous societies with a plurality of moral norms and values. Heterogeneous characteristics are most prominent in Switzerland and the Netherlands, due to the old structures of pillarization. Additionally, in all three countries the interactive legislative approach has been (indirectly) discussed in the academic field as well as in legal politics.[33] Furthermore, in Denmark, cooperation between various research institutes in the fields of law, technology and ethics has led to a set of guidelines for moral and legal decision-making that have characteristics of an interactive approach such as broad participation and the acknowledgement of moral values.[34]

1.4.3 Outline

This book is divided into three parts. Part I presents a theoretical perspective and discusses the ideal-typical model of the interactive legislative approach. Part II elaborates the case studies and their outcomes. Part III assesses the theoretical implications of the case studies for the design of a general theory of the interactive legislative approach.

Part I consists of two chapters. First, Chapter 2 presents a framework on which the quality of law may depend. This framework captures various views and perspectives about law and its quality. These dimensions are explicated in terms of validity, political morality, and fit. Second, Chapter 3 elaborates the various ideas about the communicative dimensions of legislative approaches and explores the three lines of research on the interactive

---

33 For example, Grotefeld, 'Wie wird Moral ins Recht gesetzt?', pp. 299-317; J.D. Rendtorff and P. Kemp, *Basic Ethical Principles in European Bioethics and Biolaw, Vol. 1 Autonomy, Dignity, Integrity and Vulnerability*, Barcelona, Impremta Balona, 2000; J. Rotmahr Hermann, 'Are you a Man or a Mouse?', in *Journal of International Biotechnology Law*, 2008, pp. 16-19. Their ideas are not further included in the scholarly exploration of the interactive approach as they do not present claims or explicit thoughts on the characteristic elements of the interactive approach.

34 The Danish Ministry for Trade and Industry, *An Ethical Foundation For Genetic Engineering*, Denmark, 2000.

legislative approach. The chapter then presents an initial ideal-typical model that consists of various descriptive claims.

Part II presents the outcomes of the case studies and draws some first conclusions. Chapter 4 describes the structure and approach of the case studies. Chapters 5, 6, and 7 discuss the Danish, Swiss, and Dutch case studies, respectively. In general, all three case-study countries established an interactive legislative process. However, the dynamic process faced difficulties, albeit for different reasons in each country. Chapter 8 reflects on the outcomes of the case studies and their implications for the ideal-typical model of the interactive legislative approach.

Part III elaborates some radical conceptual changes to the ideal-typical model. It focuses on two of the primary challenges to the dynamic aspect of the interactive legislative approach: the focus on aiming for consensus, which has had the effect of counteracting further norm development, and the limits of the legalistic context in which ethics committees have to operate. Chapter 9 reconsiders the concept of consensus. It argues that consensus should no longer be a focal point for legislative approaches aiming to address issues characterized by value pluralism through the use of continually developed open norms. Chapter 10 explores an alternative method of legal reasoning: an ethos of controversies. Instead of a focus on consensus or commonalities among various viewpoints, the controversies should be acknowledged. An ethos of controversies can contribute to awareness of the political nature of decision-making.

1.4.4 Methodology

The methodology used for this research is threefold. It involves literature studies, comparative analysis, and document analysis. Part I mainly involves literature studies. Studying primary and secondary literature contributed to developing the framework on the quality of law and the ideal-typical model of the interactive legislative approach.

Part II employed mainly comparative analysis and document analysis. Studying animal biotechnology regulations in the case-study countries required an analysis of secondary literature and an analysis of parliamentary documents. These analyzes were complemented by interviews with key figures.[35] Several members of ethics committees, government employees, and academics specialising in law and animal ethics were interviewed. The methodology for the case studies will be further elaborated in Chapter 4, which explores the framework for the case studies.

Part III presents a combination of both comparative analysis and literature studies. Based on the outcomes of the case studies, the model interactive approach is refined and reconstructed. Refining and reconstructing the model also involves an exploration of the theoretical implications of the outcomes of the case studies. These literature studies also contribute to the formulation of recommendations for improving the practical value of the ideal-typical model of the interactive legislative approach.

35 See Appendix I for an overview of the interviewees.

2 A Comprehensive Framework on the Quality of Law

2.1 Introduction

Reconstructing and evaluating a comprehensive theoretical interactive model of law requires an analytical framework. This chapter develops such a framework. In order to do so, it distinguishes three dimensions on which the quality of law may depend: fit, political morality, and legal validity. These dimensions are elaborated through a discussion of the characteristics of legal theories, as well as explanations of quality of law in legislative practice.

This framework will, first of all, be used to explain some of the themes that function as a basis for developing the ideal-typical model of the interactive legislative approach in Chapter 3. More importantly, however, the framework provides a tool that will be used to pinpoint, understand, and evaluate the ideal-typical model's failures, shortcomings, and differentiations in Chapters 8, 9, and 10. The framework elaborates several sub-dimensions under which differences in viewpoints about the quality of law can be defined. Among other elements, this framework distinguishes coping with problems, context, popular legitimacy, and substantive justification. This framework is extensively detailed and organized in such a way that it can provide insights into the characteristics of the interactive approach. Additionally, it allows the differences in views concerning what constitutes a qualitatively 'good' law to be identified more clearly. In this framework, 'law' refers to (the development of) legal norms in a regulatory framework.

The point of departure for this analysis is Dworkin's discussion of judicial decision-making in which he presents an argument for understanding the concept of law as integrity. This chapter starts with a short introduction of the general framework and Dworkin's analysis. The remainder of the chapter discusses the various (sub-) dimensions of the framework more thoroughly.

2.2 The Framework

Dworkin's point of departure is judicial decision-making. For him, judicial decision-making is not just a matter of applying the legal rules, but an interpretive practice.[1] In order to understand the integrity of law, Dworkin distinguishes two dimensions that contribute judicial interpretation: 'fit' and 'value'. In a nutshell, the dimension of 'fit' refers to a law's

1 R. Dworkin, *Law's Empire*, Harvard, Harvard University Press, 1993, Ch. 2.

consistency with the norms of a legal system. The law as used in court decisions must be interpreted in light of previous judicial decisions, statutes, and general legal principles. The notion of 'value' relates to interpreting the law with regard to political morality, and is therefore often referred to as the dimension of political morality.[2]

Dworkin distinguishes 'integrity' as a third conception of law. According to Dworkin, integrity identifies law as 'flowing from' procedural fairness as well as from a moral justification for the communities' political power.[3] The interpretive practice here relates to consistency with past decisions. Law as integrity understands consistency with past decisions as flowing not only from the fact that laws can explicitly be found in those past decisions, but also from the fact that laws must be consistent with the principles of personal and political morality that justify those decisions.[4] Instead of merely backward- or forward-looking interpretations, law as integrity combines both elements. Judges conceive of law as integrity because they identify legal norms as having to 'fit' with past decisions and be 'valued' as being the most fair or just decision in a political society. According to Dworkin, law as integrity is 'the best interpretation of what lawyers, law teachers, and judges actually do and much of what they say'.[5]

The concept of law as integrity contributes to understanding how judges identify law. With the construction of a framework for the quality of law, this research aims to come to a better understanding of law as well. However, in contrast to Dworkin's ideas, it focuses on understanding the differences in views on what constitutes a qualitatively 'good' law. Whereas the dimensions that Dworkin distinguishes are required for identifying law from the judge's perspective, the dimensions in the framework for the quality of law presented here are dimensions that characterize the discussion of what is required to constitute a qualitatively 'good' law. The framework does not intend to distinguish dimensions and present them in a lexical order as if they were all required to constitute qualitatively 'good' law. Rather, the framework is neutral with regard to the various normative positions about what 'good' law should be. It therefore includes all dimensions that (dependent on which normative position is defended) may be relevant for understanding what is considered to constitute and characterize a qualitatively 'good' law.

Dworkin's conception of law as integrity is, nevertheless, an interesting starting point. Similar to Dworkin, this framework intends to show that quality of law does not necessarily depend only on following the official rules. There is 'something' more to describing the quality of law in society.

Therefore, drawing on Dworkin's ideas, both 'fit' and 'political morality' are incorporated in the framework. In the following sections, these dimensions will be further concretized, modified, and elaborated.

Additionally, the framework includes a third dimension of 'legal validity'. Each legal system acknowledges official procedures that need to be followed in order for law to be

---

2 Dworkin, *Law's Empire*.
3 *Ibid.*, pp. 95-96.
4 *Ibid.*, pp. 95-96; Ch. 7.
5 *Ibid.*, p. 226.

constituted. Furthermore, some legal theories develop formal criteria that decide whether legal norms can be formally identified as such. This notion of 'legal validity' will be explored further in Section 2.5.

2.3 The Dimension of Fit

The *dimension of fit* consists of three sub-dimensions: 'fit with law', 'fit with context', and 'coping with problems'. These sub-dimensions define the background requirements that a particular legal system may require to ensure the quality of law.

2.3.1 *Fit with Law*

The sub-dimension 'fit with law' is derived from Dworkin's interpretation of 'fit'. The quality of a law can be explained in terms of consistency with other laws, legal principles and previous legal decisions. Legal norms, then, have to be construed and interpreted in a manner consistent with other existing legal norms. Inconsistency with other legal norms may decrease the quality of a law. In other words, legal norms need to be coherent with the norms and principles of the legal system. 'Fit with law' can be subdivided into two dimensions: 'sources of law' and 'legality'.

Fit with sources of law refers to a law's correlation with the sources of law that identify a system's legal norms. These sources can be traditional: statutes, the constitution, case law, treaties and customary law. According to some legal theorists, sources of law can also be legal principles that contribute to the identification of legal norms. Third, both in some jurisdictions and in international law, legal doctrine is also explicitly recognized as a source of law.

Both legal norms and the guidelines and conditions of the legal discourse in which norms are developed are derived from these sources of law. Which sources of law are most prominent and how 'law' is interpreted depends on the legal system and its legal traditions. The legal system and its legal traditions also influence whether legal principles are considered as part of the 'law' in general or whether these principles are only considered as part of the 'law' if they are incorporated into traditional sources of law.[6]

'Fit with law' can also be explained as legality. 'Legality' refers to the idea that the quality of law depends on criteria regarding the legality of legal norms. Some legal theorists identify law by means of standards that need to be followed to ensure the quality of law. For example, Fuller has distinguished various standards that reflect law's inner morality.[7]

6 The Hart-Dworkin debate as well as the debate between strong and weak legal positivists exemplifies the differences in interpretation of law. The role of legal principles is a core subject in these debates. For a comprehensive overview, See B. Leiter, 'Beyond the Hart/Dworkin Debate: The Methodology Problem in Jurisprudence', in *The American Journal of Jurisprudence*, Vol. 48, 2003, pp. 17-51.

7 L.L. Fuller, *The Morality of Law*, New Haven, Yale University Press, 1964, Ch. 4.

If law does not adjust to these standards, the actors involved will ignore or deny the legal norms. According to Fuller, law would lack quality in terms of legality if these standards were not followed. Similar standards for designing legal norms are also acknowledged in legislative practice. For example, in the Netherlands a nota 'Zicht op Wetgeving'[8] describes various standardized criteria for guaranteeing the quality of law, such as consistency, equality, proportionality, efficiency, etc.

2.3.2 Fit with Context

In addition to 'fit with law', the quality of legal norms can also be guaranteed by their fit with their context. Context, here, refers to the community in which a legal system functions. This context comprises two elements: 'political culture' and 'social context'. Quality of law can be explained in terms of 'fit with political culture', which refers to both a society's legislative culture and its political traditions. Legal norms that contradict a society's political traditions and legislative culture will, then, most likely be ineffective or even be counterproductive. For example, a regulation that channels objections to licensing decisions through official procedures on the national level despite the fact that the state is highly decentralized will likely be less effective than a more decentralized procedure. Such a regulation would enlarge the gap between citizens and politicians.

'Fit with political culture' can also be understood from an internal perspective. However, coherence with a country's internal standards regarding legislative culture and political traditions does not necessarily guarantee the quality of the political culture from an external perspective. The internal political culture could still be a morally unjustified one, such as a tyranny. From an external perspective, a 'law' may not be justified if it reflects internal but morally unjustified standards. Therefore, from an external perspective, the quality of law cannot be based only on a fit with political culture.

In addition, the quality of law may also depend on the extent of fit with their social context. 'Fit with social context' is best explained in terms of input from society in developing legal norms. 'Social context' derives from the internal perspective of society and refers to characteristic elements such as demography, cultural traditions, economic system, area, etc. According to some authors, legal norms that are developed without any input from society will most likely not adjust to the demands of society and its problems.[9] The definition of a problem, which functions as a point of departure for developing legal norms, starts with its social context. The quality of a law is, therefore, strongly linked with its fit to its social context.

---

8 Zicht op Wetgeving, 2003, WODC.

9 See for example ideas about responsive law, P. Nonet and P. Selznick, *Law and Society in Transition. Toward Responsive Law*, New Brunswick, NJ, Transaction Publishers, 2001.

2.3.3 *Coping with Problems*

The third sub-dimension of fit refers to a legal norm's capacity to coping with both existing and as-yet-unknown legal problems. At first sight, 'coping with problems' seems to intertwine with the dimension of 'fit with social context' since the social context influences the characterisation of a legal problem and the responsiveness of its approach. However, 'coping with problems' applies a different perspective. 'Social context' implies input from society and relates to the norm-addressee. 'Coping with problems' should be seen from the perspective of the legislatures who have to decide which legislative approach to follow and what kind of norms to develop in order to adequately cope with the issue at stake. Each legal problem requires a different approach and consequently a different set of legal norms to guide the behaviour of citizens.

In general, legal norms attempt to solve problems by directing citizens' behaviour. Fuller's interactionist perspective describes this as the functionalist conception of law: law can be identified as such only if it is, in its essence, focused on guiding behaviour.[10] However, coping with problems can go further than guiding behaviour. Adequately coping with both existing and as-yet-unknown legal problems may go beyond law's capacity to guide behaviour, and additionally involve a change of attitude. Law, then, can have functions in addition to or instead of the traditional instrumental and constitutional ones. In order to understand the quality of law, we should therefore focus on these additional functions of law that imply different approaches to coping with the problems at stake.

This section distinguishes between two approaches for 'coping with problems': 'coping with problems in the long run' and 'coping with the problems in the short run'. The strategic choice of either long-term or short-term coping with problems may influence a society's understanding of the quality of law. Issues like the economic crisis that recently took place in most Western countries require adequate guides for action in the short run. Some theorists argue, however, that issues such as 'unstructured problems'[11] or intractable disagreements require longer-term solutions that go beyond guiding action. Instead, coping with these complex issues requires a learning process, a change of attitude, as well as further development of legal norms. In cases like these, regulation may have additional symbolic or interactive functions.

2.4 The Dimension of Political Morality

Another dimension in this framework for understanding the quality of law is the *dimension of political morality*. Political morality, in contrast to the *dimension of fit*, is more explicitly related to the legitimacy of legal norms. The dimension of political morality consists of two sub-dimensions: 'substantive justification' and 'popular legitimacy'.

10 Fuller, *The Morality of Law*.
11 Hisschemöller and Hoppe, 'Coping with Intractable Controversies', pp. 40-60.

The dimension of political morality draws on Dworkin's dimension of 'value' or 'political morality'. Its two subdivisions derive from Van der Burg's reading of Dworkin.[12] Van der Burg claims that Dworkin's dimension of political morality is ambiguous; he breaks it down into two distinct dimensions: 'quality of law' and 'appeal and legitimacy to the population or group'.

These two sub-dimensions are both strongly related to critical as well as public morality.[13] Critical morality is the critical side of a community's moral judgment, and consists of various views, ideals, and principles on what public morality should be.[14] Public morality as a shared set of moral values can provide popular legitimacy, while reflecting on critical morality ensures that the content of legal norms is constantly challenged and can therefore be substantively justified. These dimensions therefore require distinction. That does not mean, however, that they should be strictly separated, since they interact with each other and *together* constitute political morality.

'Political morality' encompasses more than just the interaction between critical and public morality. Additional characteristics of political morality, such as policy, also affect the quality of law. To come to a better understanding of the quality of law, the dimension of political morality should not be restricted to an interpretation of critical and public morality. Instead, it must leave room for incorporating additional characteristics that are identified in legal theory and legislative practice.

As a result, this framework divides the dimension of political morality into the sub-dimensions of 'substantive justification' and 'popular legitimacy'. The remainder of this section will discuss the reading of these sub-dimensions.

2.4.1 Substantive Justification

'Political morality' embodies, first of all, 'substantive justification'. The content of legal norms requires justification in light of political morality. In legal theory and legislative practice we can identify two ways in which substantive justification can be established: with reference to 'critical morality', or by 'policies'. These aspects require further explanation.

Dworkin recognizes two arguments that can generate judicial decisions: an argument of principle and an argument of policy. Arguments of principle refer to critical morality as a source of principles and ideals. While principles and ideals are usually discussed in the context of interpretive and natural law theories, Honoré argues that critical morality

---

12 W. van der Burg, 'The Importance of Ideals', in *The Journal of Value Inquiry*, 31, 1997, pp. 23-37.

13 One remark must be made. I refer here to public morality instead of positive morality. Public morality in my opinion is a more suitable term, since positive morality misleadingly associates to positive law. In the strictest sense positive law refers to legal rules that are enacted and adopted by the proper authority. The first association with positive morality could, therefore, be that positive morality is morality enacted and adopted by the proper authority. But in reality, positive morality refers to a morality embedded in society, and can therefore better be explained as a morality of the public or in other words as a public morality.

14 F.W.A. Brom, 'Dynamische publieke moraal. Idealen als perspectief voor moraalontwikkeling', in W. van der Burg and F.W.A. Brom (eds.), *Over Idealen*, Deventer, W.E.J. Tjeenk Willink, 1998, p. 76.

applies to every law in every system. Positive law has to address critical morality during the interpretation and application of the law.[15]

Dworkin contends that arguments of policy cannot provide substantive justification in terms of political morality.[16] However, Dworkin's ideas apply only for judicial decision-making. In the broader context of law-making, ideas about what constitutes a 'good' law are not necessarily limited to substantive justification based on principles and ideals. Furthermore, in this broader context the lawmakers as elected and responsible officials come from parliament, which in reality often bases its decisions on compromises among individual goals and purposes. In legislative practice, therefore, legal norms are also justified by policy. Regulation is often used as an instrument to realize policy aims and, consequently, policies are often put forward as an argument for substantive justification of legal norms. As Waldron argues, legislatures are political institutions, and consequently legislation is strongly influenced by politics focused on establishing a majority.[17] In order to understand the quality of law, therefore, we should also analyze the legislature's policy purposes. At the same time, we should be aware that these policies require justification as well.

2.4.2 Popular Legitimacy

The second sub-dimension of political morality is 'popular legitimacy'. Popular legitimacy depends on the support of the public. There are a variety of methods of establishing popular legitimacy. However, two elements appear essential: popular legitimacy based on 'public morality' and popular legitimacy based on 'popular views'.

To explain popular legitimacy based on public morality, we can refer to Fuller's interpretation of human association. Fuller distinguishes two principles of human association: 'shared commitment' and 'legal principle'. 'Shared commitment' refers to holding associations together by shared ends, aspirations, or shared inner convictions. 'Legal principle' involves holding an association together based on the principle of reciprocity, and functions via formal rules of duty and entitlement. Both principles are correlated with and reinforce one another, but at the same time stand in a relation of polarity. Fuller argues that although 'legal principle' sets the context of legislation, 'shared commitment' is also required for a social order to function. The adequacy of a legal system depends not only on individual acts, but also on moral support and inner conviction.[18] Fuller argues that associational problems in the legal context cannot be solved by reference to formal procedures alone. 'Shared commitment' is essential as well.[19]

---

15 T. Honoré, 'The Necessary Connection between Law and Morality', in *Oxford Legal Studies*, 22, 2002, pp. 489-495.

16 R. Dworkin, *Taking Rights Seriously*, Harvard, Harvard University Press, 1978, Ch. 4.

17 J. Waldron, *Law and Disagreement*, Oxford, Clarendon Press, 1999, pp. 23-43.

18 L.L. Fuller, *The Principles of Social Order*, Selected Essays of Lon L. Fuller, edited with an introduction by K.I Winston, Oxford, Hart Publishing, 2001, pp. 96-97.

19 Fuller, *The Principles of Social Order*, p. 96.

'Shared commitment' in terms of moral support and inner conviction represents public morality. The people share their inner convictions. If norms are consistent with public morality, the people will tend to follow these norms out of a shared inner conviction. At the same time, in pluralistic societies, norms that build on shared inner convictions or shared ends are difficult to identify. Instead, actors will dissent on convictions as well as on ends. How, then, can we define conceptions of shared commitment? How should we understand popular legitimacy? To better understand how the quality of law can be identified, the framework on the quality of law needs to leave room for these questions.

The need for popular legitimacy and its methods can also be related to a society's democratic traditions. Democratic societies are founded on the ideal that the people impacted by political decisions should have influence and involvement in the decision-making process. In general, ideals of public involvement entail the concept of a people's 'sovereignty'. The extent of popular influence and the background of the people's involvement in terms of approval, acceptance or actually having a say, are related to a society's democratic traditions.[20]

Popular legitimacy is identified by either public morality or popular views. Legal theorists explain popular legitimacy in various ways, for example in terms of approval, acceptance, or actually having a say. These are, in legislative practice, translated into methods of establishing public support. In different democratic traditions, the extent of public support may depend on consistency with public morality, but does not necessarily have to. In reality, public morality is not always easy or even possible to define, especially in pluralistic societies. People may have different moral convictions, which are reflected in the difficulties of establishing a unitary public morality. At the same time, people with different backgrounds may share the same view on a particular issue.

2.5 The Dimension of Legal Validity

The final dimension in this framework on the quality of law is the *dimension of legal validity*. Each legal system and legal theory acknowledges formal rules that need to be followed in order for legal rules to be constituted. The extent, the role, and the character of these formal rules differ from one legal system and legal theory to the next. To come to a better understanding of quality of law in terms of legal validity, this dimension is divided into two sub-dimensions: 'following official procedures', and 'formal legitimacy'.

In legislative practice, law is identified as such when it takes the form prescribed by official legislative procedures. In all democratic societies, institutions must follow official

20 Views on what constitutes popular legitimacy vary in the diverse democratic traditions. Direct democracies, for example, are consistent with an explanation that defines popular legitimacy in terms of citizens' 'approval' of legal norms and decisions. This need for 'approval' is institutionalized in methods of direct involvement in decision-making that give the public an official say in the decision-making process. In representative, consensual and deliberative democracies, on the other hand, popular legitimacy is understood as the 'acceptance' of legal norms and decisions.

legislative procedures in order to ensure the validity of legal norms. The practical question of whether a legal norm is valid obviously differs from the theoretical question of what constitutes a qualitatively 'good' law.

However, validity and quality cannot be completely separated from one another. Legislative procedures are in the first place developed to protect democratic ideals and political rights. But these procedures also guarantee a level of quality in legislation. For example, a substantive judicial review that assesses the consistency of laws with the constitution and/or international treaties may ensure the validity as well as the quality of legal norms in democratic societies.

Furthermore, in some legal theories law is identified by its consistency with principles of formal legitimacy. Especially in theories of legal positivism, formal criteria are used to identify the validity of legal norms. Variations of those formal criteria include Kelsen's basic norm or Hart's rule of recognition.[21] If a legal norm reflects either a basic norm or a rule of recognition, it can be identified as law. These formal criteria also relate to the quality of law, as they ensure the validity of legal norms. Furthermore, formal criteria such as the rule of recognition arise out of a convention among officials who accept the rule's criteria as standards that govern their actions. Therefore, to understand the various views on the quality of law, we should include this dimension as well.

2.6 THE FRAMEWORK

The framework for analysing the quality of law that results from the previous sections can be summarized as follows:

Quality of law is determined by:

1. dimension of fit
   a. fit with law
      i. sources of law
      ii. legality
   b. fit with context
      i. political culture
      ii. social context
   c. coping with problems
      i. coping with problems in the long run
      ii. coping with problems in the short run
2. dimension of political morality
   a. substantive justification
      i. critical morality
      ii. policies

21 H. Kelsen, *Pure Theory of Law*, translated by M. Knight, Berkeley, University of California Press, 1967; H.L.A. Hart, *The Concept of Law*, New York, Oxford University Press, 1961 (revisited 1994).

   b. popular legitimacy
      i. popular morality
      ii. popular views
3. dimension of legal validity
   a. following official procedures
   b. formal legitimacy

2.7 Concluding Remarks

This chapter has outlined a comprehensive framework on the quality of law. It detailed three dimensions on which the quality of legal norms may depend: fit, political morality, and legal validity. This set of standards will be used to reconstruct and evaluate the interactive model of law.

Chapter 3 will use this framework to explain and develop some claims relating to the ideal-typical model of the interactive legislative approach. This more comprehensive framework with extended dimensions is required for understanding and evaluating the interactive legislative approach. Part III focuses on evaluating and reconstructing the interactive legislative approach. It identifies various failures and difficulties that the interactive legislative approach has faced in legislative practice, and evaluates these in light of the comprehensive framework developed here.

However, the framework may also have a broader application in the legal context. Although the development of this framework on the quality of law had a predetermined aim and began from Dworkin's position, it may also contribute to a better understanding of the debates in legal theory about the quality of law. Furthermore, it may provide guidelines for reconstructing and evaluating other theoretical models of law.

Nevertheless, this research presents more modest claims about the framework and its application to the reconstruction and evaluation of the interactive model of law. Additional applications would require further analysis of this framework that go beyond the aims of this research.

3 The Interactive Legislative Approach

3.1 Introduction

The aim of this chapter is to reconstruct an ideal-typical model of the interactive legislative approach. It will do so by exploring various lines of research that focus on communicative processes of law. An interactive legislative approach is an approach to developing legal norms that relies on an interactive process in which both private and public actors can participate equally and on a horizontal level. In recent years, some countries have begun to employ interactive or communicative approaches, especially in context of legislating difficult moral and political issues. Several legal scholars have explored the movement towards more interactive norm development processes in national and international contexts and have translated this change into various descriptive and normative claims about the advantages of an interactive or communicative process, its legislative functions, the nature of interaction on the horizontal level, and the need for a continuous process of norm development. However, there is not yet one ideal-typical model of the interactive approach. Therefore, further development is required.

To begin with, this chapter will elaborate on the theoretical foundations of communicative or interactive approaches in general. Communicative approaches are approaches to the study of law that focus on its communicative processes. Most approaches find their roots in the work of Fuller and Selznick.

Section 3.3 will then explore and compare three lines of research on the interactive approach to law. All these lines share an interest in the communicative processes of law. The first of these lines of research is given by Van der Burg and Brom. Van der Burg and Brom define 'interaction' and 'dynamics' as characteristic elements of an interactive approach to law.[1] They begin their analysis by pointing out that when regulating complex issues,

1 The ideas of Van der Burg and Brom must be understood in a greater movement in the Netherlands, often referred to as the communicative turn. The 'communicative turn' arose in response to the regulatory crisis. The regulatory crisis described a situation in which laws and regulations were no longer deemed effective for producing the intended outcomes. The overabundance of legal rules did not contribute to the quality of law and, consequently, their usefulness and effectiveness were doubted. As Witteveen argues, to unlock the crisis, institutions in both the political context and the legal community sought answers in *communication as the key variable*. In the Netherlands the communicative turn led to the development of various lines of research, in particular to studies on the communicative style of legislation and studies on the interactive legislative approach. These lines of research arose from different perspectives and therefore follow different focus-points. However, in context of this research the Dutch debate and its nuances are less relevant. I will therefore restrict to the ideas of Van der Burg and Brom as they emphasize the continuous process of norm-development. A characteristic element that is also of main importance in the international context. For further information on the Dutch communicative turn I refer to W. Witteveen, 'Turning to Communication in the Study of Law', in N. Zeegers, W. Witteveen, and B. van Klink (eds.), *Social and Symbolic effects of Legislation under the Rule of Law*, Lampeter, Ceredigion, Wales, The Edwin Mellen Press, 2005.

traditional regulatory approaches are often inadequate. Instead they see the emergence of an interactive paradigm, with a focus on communicative processes and a need for more flexibility.[2] Interaction can best be explained as decision-making on the horizontal level. Dynamics, in its turn, refers to a continuous process of norm development. Van der Burg and Brom's hypotheses on interaction and dynamics – which also appear in other models discussed in this chapter – will function in Section 3.4 as the starting point for the development of the ideal-typical model.

These ideas are a good starting-point for this research because they begin by assuming a change of legal paradigm when regulating complex issues with strong moral impacts – the same type of issues that this research takes as its example. However, Van der Burg and Brom have not concretized a clear concept of the interactive legislative approach or defined an ideal-typical model with clear directives. Exploration and comparison of other lines of research in the field of interactive approaches to law may therefore assist the project of reconstructing an ideal-typical model with clear directives and characteristics.

One line of research that will be used for comparison is the concept of 'regulation as facilitation' developed by Julia Black.[3] Black's ideas are of interest because her concept of regulation also focuses on issues similar to those under discussion here: regulating gene technology. Black has analyzed the regulation of gene technology issues in the UK, which in general are addressed by a more technocratic approach in which scientists are dominant in decision-making. She argues for a different view of law, which she terms 'regulation as facilitation'. As technology issues are not only given by expert opinions, but also by the concerns and opinions of lay people, the legislator needs to operate in such way that all viewpoints are considered within decision-making processes. The function of law is not to control the risks, but to facilitate the integration of all viewpoints concerning a particular technology. In this light, as will be discussed below, Black emphasizes the additional functions of law such as the cognitive and communicative functions.

The third line of research that will be examined is Brunnee and Toope's concept of the interactive international theory of law. Brunnee and Toope have developed a pragmatic theory of how international law is made. This theory emphasizes the need for interaction to sustain legality of legal norms. Brunnee and Toope's are inspired by Fuller's interactional theory. Their work shares a similar theoretical background with that of Van der Burg and Brom, and is interesting for developing the ideal-typical model because it embodies ideas on legal obligation, the need for interaction and continuous norm development.

By studying these three lines of research in both domestic and international law, this chapter hopes to provide a convincing but modest set of claims about how to engage with complex moral issues in a legal context. Therefore, Section 3.4, building on these lines of research, will flesh out the analysis of the interactive legislative approach by developing various assumptions and hypotheses explicating the two of its most prominent characteristics: interactivity and dynamism.

---

2 Van der Burg and Brom, 'Legislation on Ethical Issues', pp. 58-59.

3 J. Black, 'Regulation as Facilitation'.

3.2 Theoretical Backgrounds

This chapter singles out three primary lines of research for developing an ideal-typical model of the interactive legislative approach. However, it also seeks to contextualize these lines of research within the broader field of interactionism. In order to do so, it explores Selznick's ideas on responsive law and Fuller's ideas on interactionism. These ideas inspired the two lines of research related to domestic regulations, but also reflect the communicative reorientation of law in the international context.[4]

Selznick's ideas about responsive law have been influential in the development of interactive legislative approaches. Selznick's responsive law approach focuses on dialogue instead of forced acceptance of the rules. Legislation developed through dialogue with society responds better to that society's needs and aspirations.[5] Selznick therefore argues that rulemaking should focus on addressing the problem itself, rather than following formal rules that guarantee the validity of norms. 'The search for *implicit values* in rules and policies' is a distinct feature of responsive law, which implies an emphasis on more general ends instead of rules and procedures.[6] In this view, regulation reflects social problems. The responsive law approach is characterized by a focus on establishing the common good or aspiring to uphold common values. Rules and policies should be open for revision so that they may better reflect general values and principles.[7] Interaction between legal institutions and affected communities results in the development of legal rules characterized by open, flexible norms that are responsive to society's needs. According to Selznick, the ideal is to strive for a maximum of self-regulation.[8]

Fuller's interactionist legal theory is another important precursor of communicative legislative approaches.[9] His works 'The Morality of Law' and 'The Principles of Social Order' have strongly influenced legal theorists who defend communicative approaches to legislation. In his 'Human Interaction and the Law', Fuller analyzes the relationship between reciprocity and legislation. The paper's central question is whether the existence of law depends on a responsible interaction between legislature and subject. Underlying this question is the assumption that legal decision-making finds its basis and justification

4 Ideas of Fuller have inspired international legal scholars to explore its options for developing an international legal system and international law. International law is not subject of this thesis, however its ideas on an interactional theory of law exemplify similar ideas on the value of interaction. For example, see J. Brunnee and S.J. Toope, 'International Law and Constructivism: Elements of an Interactional Theory of International Law', in *Columbia Journal of Transnational Law*, 39/19, 2000-2001, pp. 19-74.

5 P. Nonet and P. Selznick, *Law and Society in Transition. Toward Responsive Law*, New Brunswick, NJ, Transaction Publishers, 2001, p. 14.

6 Nonet and Selznick, *Law and Society in Transition*, p. 79.

7 P. Selznick, *The Moral Commonwealth. Social Theory and the Promise of Community*, Berkeley, CA, University of California, 1992, p. 338.

8 P. Selznick, *The Moral Commonwealth*, p. 470.

9 For example B.M.J. van Klink and W.J. Witteveen, 'Why is Soft Law Really Law?', in *RegelMaat*, (3), 1999, pp. 120-140; Brunnee and Toope, *Legitimacy and Legality in International Law*; Van der Burg and Brom, 'Legislation on Ethical Issues'.

ultimately in human interaction. Fuller argues that in order to be effective, legal systems should adhere to the principles of 'reciprocity' and 'co-operative effort'.[10]

Fuller relates 'reciprocity' to human associations, as legal rules are one of the elements that hold human associations together. The principle of reciprocity entails a morality of duty. Duties lay down the basic rules for a social order.[11] As a minimum, citizens have a duty to follow the basic rules that keep associations together. This is a sort of 'vertical' interaction, as it flows between an association and its members.

A second principle that holds human associations together is the principle of 'shared commitment'. The principle of shared commitment can best be explained in terms of a search for common ends or purposes, as the legal principle 'refers to the situation where an association is held together and enabled to function by formal rules'.[12] This is a sort of 'horizontal' interaction, as it flows between the members of an association. Together, these principles hold human associations together and define the mode of interaction between actors in a human association. Fuller argues that a responsible interaction depends on both principles.[13]

In communicative legislative approaches, interaction implies a form of communication between legislature and subject. This interaction has both a vertical and a horizontal structure. An interactive practice is defined as a process of norm development in which all actors communicate equally on a horizontal level. In order to ensure responsible interaction that can stimulate ongoing norm development, interactionists argue that a legal system should adopt horizontal structures of norm development. If there are only vertical relations between legislature and subject, the effectiveness of legal rules is dominated by a morality of duty related to a principle of reciprocity. Vertical relations must therefore be supplemented by horizontal interactions. Shared commitment depends on shared goals; without a shared commitment, a responsible, horizontal interaction between actors of the association is hard to enable. Shared goals are what define the field for interaction in terms of horizontal co-operation.

The principle of shared commitment relates to what Fuller distinguishes as a morality of aspiration. A morality of aspiration refers to values that people cherish. In communicative approaches, a morality of aspiration is important for a responsible interaction between legislature and subject, as it acts as a guideline for action. Aspiration in itself does not provide a good basis for legal rules, even though common aspirations and shared commitments to these aspirations are what hold an association together.[14] Vertical structures of interaction, which stress the element of reciprocity, are therefore additionally important.

Selznick's ideas about responsive law and Fuller's ideas about interactionism have had an important influence on research about interactive approaches. The emphasis on the communicative dimensions of law, the use of aspirational norms, and the importance of

10 Fuller, *The Morality of Law*, pp. 39-40.
11 *Ibid.*, Ch. 1.
12 Fuller, *The Principles of Social Order*, Ch. 2.
13 *Ibid.*, Ch. 1.
14 Fuller, *The Morality of Law*, Ch. 2.

interaction are all visible in both Selznick's and Fuller's theories and in the communicative legislative approach.

3.3 The Various Interactive Approaches

The aim of this research is to reconstruct an ideal-typical model of the interactive legislative approach. This ideal-typical model will additionally serve as a descriptive model for testing its practical value.

Van der Burg and Brom's ideas will function as a starting point.[15] The essential elements of Van der Burg and Brom's interactive legislative approach reflect the approach's two main elements: interaction and dynamics. The exploration and comparison of the three lines of research that are analyzed in this chapter will circle around these two main elements. Additionally, as a third element, it will discuss the functions of law.

Interaction

First, the interactive legislative approach emphasizes the need for a process of *interaction* between the legislature and society or relevant sectors within a society. Both vertical and horizontal structures of communication are relevant for guiding interactive processes.

In traditional approaches to law, vertical structures function as a basis for guiding interaction. As Brunnee and Toope claim, reciprocity is the basis for human interaction. They argue that norms are legal so long as they meet the criteria of legality: generality, promulgation, non-retroactivity, clarity, non-contradiction, not asking the impossible, constancy, and congruence between rules and official action.[16] These criteria are not, however, a checklist for guaranteeing the quality of law. According to Fuller, these criteria affect fidelity and consequently people's willingness to reason within the context of the norms. Meeting these criteria will ensure reciprocity, as people will be willing to follow the rules. Reciprocity can thus be seen as a basis for guiding human interaction.

The interactive approach, by contrast, also emphasizes horizontal processes of decision-making. The interactive legislative approach's interpretation of the role of the legislature thus involves a prominent role for Fuller's 'co-operative effort' between legislature and citizen or actors from the field. This explicitly highlights a shift from the vertical relation of conditioned self-regulation towards a horizontal level of interaction. This interpretation

15 Van der Burg and Brom in early studies of the interactive legislative approach started from various empirical claims about reported changes towards a more interactive approach. These empirical claims inspired a set of normative ones. However, in the ideal-typical model, as being a descriptive model, the normative claims will not be included. Van der Burg, 'The Irony of a Symbolic Crusade: The Debate on Opening Up Civil Marriage for Same-Sex Couples', in N. Zeegers, W. Witteveen, and B.M.J. van Klink (eds.), *Social and Symbolic Effects of Legislation Under the Rule of Law*, Lewiston, New York, The Edwin Mellen Press, 2005, pp. 249-276.

16 Brunnee and Toope, *Legitimacy and Legality in International Law*, pp. 6; 37-42.

of interaction breaks out of traditional paradigms. Similar developments can be noticed in the ideas of Brunnee and Toope. Although they emphasize the importance of reciprocity as a basis for guiding human interaction, Brunnee and Toope also consider horizontal interaction as equally important for establishing both an international legal system and international legal norms. According to them, focusing on the horizontal level of interaction is particularly appealing when dealing with issues that cut across state borders.[17]

The additional focus on co-operative effort gives the interactive process a number of advantages for coping with complex moral issues. Studies of the interactive legislative approach emphasize that these advantages include an increase in citizens' acceptance of legal rules. In modern pluralist democratic societies, citizens do not want to be forced by law to change their convictions and ideals. Since moral issues confront citizens' moral convictions and ideals, involving people in the development of the legal rules makes them easier to accept. Furthermore, the interactive process leaves more room in the legal framework for applying and adapting legal rules to concrete situations. After all, moral convictions are also given substance in concrete situations. With the help of the interactive approach, the legislature can gain a more comprehensive understanding of existing practice and ethics as it interacts with the actors on the ground. This involves both allowing more expert participation in decision-making as well as broader public participation. The legislature cannot capture all (moral) concerns and viewpoints, and should therefore consult both experts and lay people for decision-making.[18] Additionally, Grotefeld argues that a horizontal and inclusive process of decision-making improves political legitimacy.[19]

Van der Burg and Brom argue that moral issues are regulated through an interactive process in which the legislature and (relevant sectors of) society co-operate on a horizontal level.[20] This emphasis on interaction and horizontal cooperation makes the interactive legislative approach 'the best method to remain oriented to the ideals and the values that the law aims to realize'.[21]

Black also emphasizes the need for interaction, which she refers to as negotiation and integration. Similar to Van der Burg and Brom, Black starts from the issues that need to be addressed. Black distinguishes three approaches for regulating those technology issues that she defines as 'intractable problems'. The number of different conceptualisations of the issue and of its solutions makes a technology like, for example, gene technology difficult to regulate. The most prominent view on how to address such issues is what is termed

17 Brunnee and Toope, 'International Law and Constructivism', pp. 43-61; Although they do not explicitly refer to an interactive model of decision-making, earlier roots can be found in the transnational legal process as described by Koh in the 1994 Roscoe Pound lecture. Here, Koh presents similar claims on the value of interaction; especially for its dynamic and normative features. See H.H. Koh, 'Transnational Legal Process', p. 184.

18 Black, 'Regulation as Facilitation', pp. 621-622.

19 S. Grotefeld, 'Self-Restraint and the Principle of Consent', in *Ethical Theory and Moral Practice,* 2000, pp. 77-92.

20 Van der Burg and Brom, 'Legislation on Ethical Issues', p. 61.

21 *Ibid.*

the 'technocratic view'.[22] In the technocratic view, scientific language is dominant. The goal of regulation is to determine risks and identify hazards from a scientific perspective. Public acceptance or public understanding of the law is perceived as being correlated with public understanding, and can be improved by means of public education. Regulation is aimed at controlling risk: any public concerns or moral issues involved are secondary.

The second approach Black describes for regulating intractable technological issues is 'modified technocracy'.[23] In this approach, the scientific viewpoint is still dominant, but there is more room for public involvement. This involvement, however, is limited to an assessment of public worries concerning the risks. Other concerns and opinions are not addressed. Pragmatic desires, such as gaining more public support, play a role here.

Black criticizes both the technocratic and modified technocratic views because they exclude important characteristic elements of the issue at stake. Regulations regarding gene technology cannot concern themselves only with scientific risks. Instead, lay perceptions should also be taken into account. As an alternative, therefore, Black presents a third view which is much more radical and exhibits several similarities with Van der Burg and Brom's interactive model: regulation as facilitation.[24] This proposition involves the rejection of the technocratic view and the dominance of science in decision-making. Instead, in the 'facilitative' approach, all viewpoints including lay perceptions are given the same weight in decision-making. Contesting the neutrality of science is a central part of the discussion. This may not be the main concern of Van der Burg and Brom's interactive model, however both positions share a concern for integrating all viewpoints, not only the scientific ones.

Dynamics

The second distinctive characteristic of the interactive legislative approach is that interaction is still required during the implementation phase: the *dynamics* of the process. Van der Burg and Brom emphasize the interplay between interaction and continuous processes of norm development. If legislation does not want to play catch-up with society, the legal framework needs to be flexible. To engage with new trends, the interactive legislative approach uses open norms and a process of continuous norm development that is sustained after the initial formation of legal norms. Interactive legislative processes facilitate both ongoing reflection on relevant moral issues and an ongoing debate. This process may stimulate the development of moral as well as legal norms. In fact, these may even reinforce each other. This presumes a continuous interaction between legal and moral norm development so that legal norms stay orientated to practice.

Similar ideas can be find in Brunnee and Toope's interactional international theory. Brunnee and Toope present a pragmatic theory consisting of three main elements.[25] They first emphasize that law can only emerge if it is built on what they call shared

22 Black, 'Regulation as Facilitation', p. 653.
23 *Ibid.*, p. 655.
24 Black, 'Regulation as Facilitation', p. 655.
25 Brunnee and Toope, *Legitimacy and Legality in International Law*, Ch. 1.

understandings. Legal norms can be seen as social norms. Consequently, legal norms are connected to social practice. Without such a connection legal norms can easily become dead letters. In order to ensure commitment to the law, therefore, norms should thus be grounded in shared understandings. But, as Brunnee and Toope argue, interactional law can only emerge if those shared understandings become fused with a 'practice of legality' rooted in Fuller's eight criteria of legality.

While meeting these eight criteria creates an initial commitment to the norms on a basis of reciprocity,[26] Brunnee and Toope's interactional theory also includes the concept of a continuing practice of legality.[27] Legal norms require constant reinforcement. Otherwise, norms run the risk of losing public commitment and becoming dead letters as they deviate from social practice. Interaction must continue in order to sustain the legal norm's influence. Continuing interaction in itself is not sufficient. A specific type of interaction is necessary: a continuous practice of *legality*. Interaction encourages the orientation of norms to shared understandings, and along with a practice of legality continuously stimulates a sense of obligation, resulting in the development of still more substantive rules and substantive commitments.

When norm development is reconceived as an ongoing process, it changes the relationship between means and ends in law. Fuller, for example, criticizes the means-ends dichotomy. Traditionally, this debate has circled around the question of whether the law should be seen as an end in itself, or as a means for reaching an extra-legal end. In studies on the communicative style of legislation, the law is often considered both an end of the legislative process as well as a means for further norm development. Fuller argues that there is no clear distinction between means and ends and emphasizes that a means for some may be an end for others. Moreover, means constitute the meaning of ends, and means and ends exist in a relation of pervasive interaction with each other.[28] This interpretation seems to lead to a better understanding of the continuous norm development process, in which ends are clarified by means of interaction.

Here, the role of consensus in horizontal decision-making becomes clear. As Van der Burg and Brom explain, the interactive paradigm emerged in response to the inadequacy of the traditional paradigm in situations where clear norms did not exist.[29] The interactive paradigm does not presuppose moral consensus, since moral consensus simply does not exist. Instead, it positions consensus as the objective of an ongoing moral debate. As mentioned before, Brunnee and Toope argue that in order for norms to be qualified as law, they need to be grounded in shared understandings. Legal norms must rest on shared understandings if they are to guide interaction. The concept of shared understandings refers to the background knowledge that is required to comprehend emerging social norms and

26 See the previous paragraph.
27 Brunnee and Toope, *Legitimacy and Legality in International Law*, p. 352.
28 P. Westerman, 'Means and Ends', in W. van der Burg and W. Witteveen (eds.), *Rediscovering Fuller: Essays on Implicit Law and Constitutional Design,* Amsterdam, Amsterdam University Press, 1999, pp. 47-62.
29 Van der Burg and Brom, 'Legislation on Ethical Issues', pp. 58-59.

practices.[30] Shared understandings are a basis from which norms can develop and from which commitment can be created.

Because of their preoccupation with shared understandings, at first sight it seems that Brunnee and Toope, unlike Van der Burg and Brom, presuppose consensus as necessary for decision-making. However, these shared understandings do not simply exist. They are generated through interaction and negotiation.[31] Brunnee and Toope explain that in international law there exists a rather limited shared understanding that law is needed. Through interaction and negotiation these shared understandings may grow. A basis is required, but it might begin as something very thin. A shared basis does not mean people need to have a common goal. It is enough to start from a shared understanding that we need a certain extent of control.[32] These shared understandings are continuously constructed and reconstructed through the practice of legality. They are both a means to guide interaction as well as an end in a sense that these understandings will be subject to further substantive development.

The search for consensus justifies a continuous process. Even though the interactive approach recognizes that interaction may not guarantee a final consensus, it at least makes this result more likely.[33]

Functions of Law

The interactive legislative approach envisions a changing role for legislation, with an emphasis on the ongoing norm development process. This does not imply that statutes have no function. However, the constitutive function of statutes is only one part of a broad range of functions of law, which extend from the conception to the implementation phase. According to Brunnee and Toope, interactional legal practices sustain legality if the focus is shifted away from the formal institutions of lawmaking, and toward the interplay between the actors operating in the legal practices.[34] It is not that formal sources such as treaties are not important at all, but a richer understanding of legality would contextualize those formal sources within the broader international legal practice. Treaties are a step in 'the continuing interactions that make, maintain, remake or even unmake international law'.[35]

We can identify four functions of law in the interactive legislative approach: the articulation function, the communication function, the interaction function, and the coordination function. In the interactive legislative approach, law's first function is the articulation of the issues at stake. Interaction between concerned actors and the legislature leads to an articulation of the various viewpoints and concerns that characterize the issue.

30 Brunnee and Toope, *Legitimacy and Legality in International Law*, p. 350.
31 *Ibid.*
32 Brunnee and Toope, *Legitimacy and Legality in International Law*, p. 80.
33 Van der Burg and Brom, 'Legislation on Ethical Issues', p. 61, p. 64.
34 Brunnee and Toope, *Legitimacy and Legality in International Law*, pp. 20-33.
35 *Ibid.*, p. 352.

Law's second function relates to the communication between the government and the actors involved. The horizontal interactive structure ensures that communication is direct and open. Moreover, interaction allows for the development of a common framework and vocabulary, which also improves communication.

Law's third and most characteristic function under this approach is the interaction function. In an interactive legislative process, the legal framework stimulates interaction between the legal context and the field of application. This leads to a dynamic interplay between legal and moral norm development. Interactive legislation can lead to the development of new frameworks of normative thinking.

The fourth and final function of law in an interactive approach is the coordination function. Institutionalisation of the interactive process and the framework for communication ensures control and procedural fairness.[36]

These functions partly overlap with the functions Black distinguishes. Black emphasizes three dimensions of law: structural, cognitive and communicative. All three are needed to facilitate integration. The first refers to the institutional context in which regulation is developed. This function can be compared with the coordination function. The cognitive dimension implies that the regulator has to take into account the perceptions of the participants on the characteristics of the issue at stake. This is similar to the articulation function. The communicative dimension refers to interaction between the participants. Here, negotiation and finally integration takes place. We can recognize here both the communicative and interaction functions of law.

Black argues that it is not the function of law to control risks, but instead to facilitate the integration of all viewpoints. Law cannot replace the variety of viewpoints with a single, scientific viewpoint. Law must instead facilitate 'the full integration of different perspectives in the regulatory process'.[37] Black uses the term 'negotiation of regulation' to indicate that the goal of integration is to arrive at solutions to intractable problems such as gene technology that are accepted by the actors involved. Black points out that the desire for integration is often seen as requiring redesign of regulatory institutions. Scholars suggest that proceduralisation, forums for consensus-building and co-operation, democratisation, and/or the adoption of communicative models of regulation will be helpful for achieving integration. Black argues, however, that while these ideas contribute to integration, they also seem to ignore an important and fundamental problem: translation. Each viewpoint has its own cognitive framework that cannot simply be replaced by a new cognitive framework and a common language. Integration and negotiation cannot be achieved without bridging this language barrier. Regulation can play the role of interpreter or translator by 'putting the views of each set of participants into a language that the others can understand'. [38]

---

36 Vorstenbosch and Ippel, 'De weg naar de wet. Wetgeving over ethische kwesties: Een interactief proces', pp. 60-66.

37 Black, 'Regulation as Facilitation', p. 622.

38 Black, 'Regulation as Facilitation', p. 623.

3.4 The Characteristics of the Interactive Legislative Approach

This section elaborates the distinctive features of the interactive legislative approach. It presents a reconstruction of an ideal-typical model of the interactive legislative approach, consisting of descriptive claims on how this approach functions in the ideal-typical model. This set consists of three main theses, five assumptions derived from these theses, and three hypotheses.

Theses

First of all, the set of distinctive features consists of two descriptive main theses that are based on two normative theses as presented by Van der Burg and Brom in their paper on the change towards an interactive paradigm.[39] They emphasize that the interactive legislative approach is a dynamic process with an interactive process of norm development. These two theses are complemented by a third thesis related to the characteristic functions of the approach. These features are elaborated as theses because of their empirical origin.[40] They are, therefore, posited as descriptive claims and form the basis of the model.

> *Thesis 1*
>
> *The process of legislation on ethical issues is structured as a process of interaction between the legislature and society or relevant sectors of society, so that the development of new moral norms and the development of new legal norms may reinforce each other.*
>
> *Thesis 2*
>
> *Legislation on ethical issues is designed in such a way that it is an effective form of communication and that, moreover, it facilitates an ongoing moral debate and ongoing reflection on those issues, because this is the best method to ensure that the practice remains oriented to the ideal and values the law tries to realise.*
>
> *Thesis 3*
>
> *In addition to traditional legislative functions, the interactive legislative approach acknowledges an articulation function, a communicative function, an interaction function, and a co-ordination function.*

---

39 In fact, these theses closely relate to the normative theses that Van der Burg and Brom posited; 'should be' in their claims is replaced by 'is' in my theses. Van der Burg and Brom, 'Legislation on Ethical Issues', p. 61.

40 Van der Burg, 'The Irony of a Symbolic Crusade', pp. 245-276; W. Van der Burg, 'Wetgeving inzake morele kwesties: Een procesgerichte benadering', in *Ars Aequi*, 47/2, 1998, pp. 79-87; Van der Burg and Brom, 'Legislation on Ethical Issues'.

Assumptions

Second, the set consists of six descriptive assumptions derived from the main theses. These assumptions concretize the main theses. They are termed 'assumptions' because they have not yet been tested empirically. These assumptions follow from a reconstruction of the various lines of research in light of the main theses.

Assumption one relates to the first main thesis and concerns the interactive character of the norm development process. The interactive character must be understood in light of the focus on co-operative effort in addition to the focus on reciprocity. Section 3.3 described various advantages of the interactive process for addressing complex moral issues, such as its tendency to increase citizens' acceptance of legal rules, to provide room for applying and adapting legal rules to concrete situations, to increase legislative understanding of the field in which the issues must be understood, and to leave room for more expert participation in decision-making.

Assumption three follows from the empirical studies of Van der Burg (1998, 2005) and Black (1998). Van der Burg and Black base their claims concerning the interactive legislative approach on the evaluation of moral and technological regulations.

Assumptions two, four, and five all reflect the dynamic character of the norm development process. Open norms are the interactive legislative approach's response to complex moral issues.[41] As Brunnee and Toope state in response to the 'soft' law debate, 'soft' norms may sometimes possess even more obligatory force than 'hard' law. After all, these norms are grounded in shared understandings and have therefore achieved fidelity.[42]

The lack of consensus on moral values and the existence of conflicting moral values that characterize these issues imply a need for open norms. Consequently, moral values play a prominent role in all stages of the exploration and ongoing development of these open norms. Ensuring moral debate in all stages of norm development implies both an explicit inclusion of (controversial) moral values in these stages and a reflection upon these moral values within these stages.

Assumption 5 relates to consensus. Consensus-building helps to ensure a continuing debate when it is perceived as a way of encouraging integration.[43] The idea of a possible consensus or an interpretation of a (temporary) decision as if it were a consensus will also stimulate this constructive interaction.[44]

> *Assumption 1*
>
> *Various social actors, including governmental bodies and citizens, are involved in a co-operative effort on a horizontal level during all phases of norm development.*

41 Van der Burg, 'The Irony of the Symbolic Crusade'.

42 Brunnee and Toope, *Legitimacy and Legality in International Law*, pp. 50-55.

43 Black, 'Regulation as Facilitation', p. 622.

44 Van der Burg and Brom, 'Legislation on Ethical Issues'.

Assumption 2

Open, aspirational norms play an important role in structuring and guiding the ongoing process of norm development. By the same token, an ongoing process of norm development stimulates the modelling of open, aspirational norms.

Assumption 3

The interactive legislative approach is domain-specific: it is best suited for addressing moral issues.

Assumption 4

Moral values related to the controversial policy issues at stake are incorporated into all stages of norm development.

Assumption 5

(Moral) consensus is not presupposed as a starting-point for norm development, but is instead regarded as one of the purposes of norm development.

Hypotheses

The final set of distinctive features presents various hypotheses. These hypotheses attempt a further concretisation of the model. However, these claims cannot be directly derived from the main theses. Instead, they are presented as descriptive statements that must be verified against practice 'on the ground'. Consequently, these claims are referred to as 'hypotheses'.

These hypotheses have been developed as a reflection of the theses and assumptions describing the framework on the quality of law elaborated in Chapter 2. Assessing political morality brings us to the first two hypotheses. A dynamic and interactive process of norm development may contribute to increase popular legitimacy. Brunnee and Toope emphasis the increase of commitment that occurs when legal norms are built on shared understandings, meet Fuller's criteria, and are embedded in a practice of legality.

At the same time, the dynamic character of the interactive approach is also a source of substantive justification, as the use of open norms requires the explication of moral values. In the interactive legislative approach, ideals, principles and policies play a prominent role in establishing the substantive justification of aspirational norms.

Van der Burg introduces a distinction between ideals, principles and policies in the sphere of Fuller's 'morality of aspiration'.[45] Policies and principles are less abstract conceptions of ideals, but still cannot be fully realized. Therefore, they fall into the sphere of aspiration. These conceptions could be used to justify aspirational norms, but could also

45 W. van der Burg, 'Morality of Aspiration', in W. Van der Burg and W. Witteveen (eds.), *Rediscovering Fuller: Essays on Implicit Law and Constitutional Design*, Amsterdam, Amsterdam University Press, 1999, pp. 176-177.

be understood as reflecting critical morality. The latter interpretation reflects the deontological approach of moving from ideals to principles.[46] Both ideals and principles can function as critical morality. However, in this context we should not ignore the substantive justification that may be found in policies. As Witteveen points out, communicative styles of legislation are motivated by ideology, but can also have pragmatic foundations.[47] For some kinds of problems, an interactive approach just works better, since clear-cut regulation is difficult to establish. On the other hand, policies may have a strong ideological foundation or may express the political identity of a society. A teleological approach to ideals will therefore lead to those policies.[48]

The interactive approach is also characterized by its emphasis on the importance of ongoing reflection and debate. Chapter 2's dimension of 'coping with problems' is particularly relevant here. Emphasising further norm development is important because norms related to complex moral problems cannot be established on the facts and values known at any one single moment. A short-term strategy would have no need for ongoing development; the status quo would do. The interactive legislative approach interprets 'coping with problems' broadly. This approach is not focused on a clear end for which the law is a means, and therefore reflects a long-run problem-solving strategy.

> *Hypothesis 1*
> *The interactive legislative approach emphasizes popular legitimacy as based on involvement of a broad spectrum of actors.*

> *Hypothesis 2*
> *The interactive legislative approach emphasizes the substantive justification of laws based on a critical reflection in light of principles, ideals, and policies.*

> *Hypothesis 3*
> *The interactive legislative approach focuses explicitly on long-run problem-solving, understood in terms of coping adequately with the characteristic issues of a problem rather than reaching a fixed aim.*

3.5 FINAL REMARKS

This chapter developed an ideal-typical model of the interactive legislative approach. The model will be used to structure the case studies in Part II as well as to evaluate the outcomes of these case studies. Chapter 4 will translate the model's assumptions and hypotheses into themes for structuring the case studies. In addition, Part III will use the results of the case studies to develop some modifications and corrections of the original model outlined here.

---

46 Van der Burg, 'Morality of Aspiration', pp. 178-179.
47 Witteveen, 'Turning to Communication in the Study of Law'.
48 Van der Burg, 'Morality of Aspiration', p. 179.

Part II

4 Introduction to the Case Studies

4.1 Introduction

The aim of Part II is to examine the practical value and design of the interactive legislative approach. It assesses the quality of law (as defined in Chapter 2) in the context of the interactive legislative approach from both an empirical and an evaluative perspective. Chapters 5, 6 and 7 deploy an empirical perspective, analyzing the regulation of animal biotechnology in the three case-study countries. Chapter 8 then takes an evaluative perspective, examining these outcomes in light of the ideal-typical interactive legislative approach. It also draws some conclusions about the practical value of the interactive legislative approach, and gives some preliminary explanations for its difficulties and shortcomings.

The exploration of the case studies will be structured around the ideal-typical model of the interactive legislative approach developed in Chapter 3. The focal points for the examination of these case studies and the presentation of their outcomes are derived from the application of the ideal-typical model to each country's regulation on animal biotechnology. As a result, the case studies do not present a comprehensive comparative analysis of legislative approaches in the three different countries, but rather involve an exploration of the interactive characteristics of the countries' legislative approaches.

Chapters 5, 6, and 7 structure their presentations of the outcomes of the case studies around several general themes regarding the claims of the ideal-typical model of the interactive legislative approach. This approach ensures a comprehensive analysis of the practical value of the interactive legislative approach. It permits the analysis to go beyond merely testing the practical value of the hypotheses and assumptions of the ideal-typical model. To that extent, a 'theme-based' analysis leaves room for identifying additional or varying concretisations and interpretations of the normative ideals of the interactive legislative approach. Moreover, structuring the analysis around general themes allows for a better understanding of the case-study countries' legislative approaches, since it is not limited to the model's assumptions and hypotheses.

Section 4.2 introduces the general themes under which the characteristics of the interactive legislative approach are analyzed: 1) public involvement in legislation; 2) the role of moral values in legislation; and 3) the legitimacy of legal norms. It explains how the descriptive claims about the ideal type fit under these themes.

Section 4.3 discusses questions concerning the conditions under which an interactive legislative approach is most likely to function adequately. It can be questioned whether the practical value of the interactive legislative approach depends on a country's political culture. The case studies will, therefore, analyze the democratic and legislative traditions of the case-study countries.

Section 4.4 concludes by describing the general layout of the chapters presenting the case studies.

4.2 Characteristics of a Legislative Approach

This section discusses the general themes that organise Part II's analysis of the selected countries' legislative approaches to animal biotechnology: the role of moral values in legislative processes, the role of the public in legislative processes, and the legitimacy of legal norms. These three themes reflect the characteristics of the interactive legislative approach, and therefore contribute to testing its practical value. Additionally, they facilitate Chapter 8's comparison of the case-study outcomes against the ideal-type of the interactive legislative approach.

Role of Moral Values in Legislation

Chapter 1 identified three characteristic issues concerning animal biotechnology: 1) a lack of knowledge about the possibilities, the potential impacts, and the consequences of this technology both at this moment and in the near future; 2) rapid changes in society and technological applications; and 3) a profound moral pluralism. Moral controversies mainly arise in the context of the second and the third characteristics. Rapid changes may lead to new moral dilemmas in which a variety of values conflict. Furthermore, the existence of a profound moral pluralism brings with it diverse and controversial interpretations of these values and disagreement about how they should play a role in legislation.

One of the assumptions of the ideal-typical model relates to the interactive legislative approach's domain-specificity. The model hypothesises that the interactive approach will function best for addressing issues with strong moral impacts. According to proponents of the interactive legislative approach, moral values should play a prominent role in regulating these issues. The model developed in Chapter 3 posits that in an interactive legislative approach, moral values play a prominent role in all stages of norm development. This claim relates to the domain-specificity as well as to the dynamics of the interactive legislative approach. As Vorstenbosch and Ippel argue, an interactive process can ensure that practice stays oriented to the ideals and values the law tries to realize.[1]

The case studies distinguish three stages of legislation in which moral values can explicitly or implicitly play a role:

1. the legislative process
2. the legal texts
3. the implementation-phase

Chapters 5, 6 and 7 will use these three stages as a structure for examining the role of moral values and for presenting the outcomes of the case studies. Moral values can play various roles in these stages. In the legislative process, moral values can be the subject of reasoning or deliberation on the issue at stake. For example, problem-definition involves

1 Vorstenbosch and Ippel, 'De weg naar de wet', pp. 49-67.

taking stock of various viewpoints, concerns, and moral values. Consequently, addressing the problem also implies addressing the moral controversies associated with the problem. In the second phase, moral values can be included in the regulatory framework or other legal texts such as the explanatory documents. The moral values are not legal norms in themselves, but frame the context for the interpretation and implementation of legal norms. Moral values should also be taken into consideration when applying, implementing and developing legal norms. The ideal-typical model assumes that the development of legal norms interacts with the crystallization of the moral values. Considering moral values during the development of the legal norms will additionally contribute to the interpretation and further definition of these moral values.

Public Involvement in Legislative Procedures

One of the purposes of this analysis is to determine whether the three countries employ interactive procedures for developing norms concerning animal biotechnology. Several of the claims set out in Chapter 3's model of the interactive legislative approach, such as theses 1, 2, and 3, assumption 1 and hypothesis 1, emphasize the importance of public involvement in the legislative process. The interactive character of public involvement implies that various actors take part in the norm development process, and, most importantly, that this participation occurs on a horizontal level and is the result of a cooperative effort in which these various actors deliberate on an equal footing. In an ideal-typical interactive legislative approach, the public and experts are actively involved.

Participation can be accomplished via various methods. The analyzes of the case studies will therefore also involve an examination of the various instruments for ensuring public involvement. They will discuss the role of experts as well as the motivations for and background of public and expert involvement.

Legitimacy of Legal Norms

The understanding of legislation and legislative approaches is also influenced by reflections on the legitimacy of legal norms. The model ideal-typical interactive legislative approach states that the quality of the (open) norms that result from interactive legislative processes depends on the involvement of a broad spectrum of actors and on the substantive justification of these norms on the basis of a critical reflection on principles, ideals and policies. The need for broad involvement of actors reflects an alternative to the traditional interpretation of popular legitimacy as public support or public acceptance. The complex structure of issues such as animal biotechnology means that public support is rather difficult to establish. Popular legitimacy must therefore rely on interaction and public involvement in norm development. In addition, legitimacy requires substantive justification in light of principles, ideals and policies.

These sources for legitimacy can be traced back to the dimension of political morality described in Chapter 2's framework on the quality of law. Chapters 5, 6, and 7 will attempt to analyze whether the case-study countries use similar justifications to ensure the quality of law. The chapters will therefore explore all sources for legitimacy that can be recognized in the legislative processes of each country's regulation on animal biotechnology.

4.3 Conditions for an Interactive Legislative Approach

In addition to featuring certain characteristics in terms of the role of moral values, public participation, and the legitimacy of legal norms, a successful interactive legislative process may depend on a country's political culture, which must be able to accommodate the interactive approach. This section addresses whether the adequacy of the interactive legislative approach indeed depends on a country's democratic and legislative traditions.

Democratic Traditions

A country's particular democratic traditions are one of the distinctive elements of its political culture. For example, different democratic traditions involve different perceptions of public involvement, and each country's democratic traditions affect its standards for the legitimacy of legal norms and influence the establishment and interpretation of institutions.

The case studies aim to achieve a comprehensive perception of the elements that characterize the democratic traditions of the case-study countries. This exploration of the democratic traditions of the three case-study countries is aimed at analysing whether they can accommodate an interactive legislative process. This analysis will contribute to answering the question of whether the adequacy of a legislative approach depends on specific democratic traditions. Furthermore, it will allow the identification of traditions that offer comparatively better conditions for an interactive process.

The case-study countries were chosen with regard to their democratic traditions. Nether; is often characterized as a deliberative democracy, whereas Switzerland and the Netherlands are said to show elements of consociational democracy. Moreover, Switzerland has developed various institutions based on the ideal of direct democracy. All of these democracies ensure broad public involvement in decision-making, although for different reasons. Choosing case studies from three different kinds of democracies offers the opportunity to explore whether the reasons for encouraging broad involvement also influence the success of interactive legislative approaches. The analysis of the democratic traditions of these three countries is carried out from both a theoretical and a practical perspective. A study of the literature in law and political science will form the basis of the theoretical perspective. The practical perspective stresses the elements of the decision-making process and other political instruments that contribute to decision making or policy-making. This dual perspective will contribute to identifying the specific conditions for a process of broad involvement, both in theory as well as in practice.

Role of the Public

The public plays a prominent role in the interactive legislative model. The public may function as co-producers of legislation, or may deliberate with decision-makers on a horizontal level. The public can also have a role outside the legal context in the political context in general. This may not imply direct involvement, but can be relevant for politics and decision-making. The case studies will focus explicitly on the role of the public in order to allow for a better understanding of its role in regulating animal biotechnology.

This exploration of the role of the public focuses on two points. First of all, it examines the role of the public in relation to the role of experts. The public is often represented by interest groups that participate in decision-making processes as experts or representatives. Furthermore, experts are often part of the field, which makes it easier to adjust legal norm development to reality. Besides, expert committees can be appointed to bridge the gap between politics and the people as well as to stimulate public debate.

Second, there are two conceptions of the public that represent two different forms of public influence: legitimating and participatory.[2] On the one hand, 'the public' can be seen as one actor; as a collective identity. This collective public can function as the legitimating force behind the whole political system. Alternatively, 'the public' can be understood as a mixture of various actors who can be represented in all kinds of organisations and expert groups.[3] The public then consists of individualized citizens with splintered interests. The various interests of a single citizen can be represented via different channels. This fragmented public influences the state through participation. It is important to acknowledge this twofold understanding of the public, since it provides insights into the various levels of public involvement noted in the case studies.

Legislative Cultures

Each country also has a distinctive legislative culture. This study is especially concerned with studying the legislative culture of each case-study country, since jurisprudential traditions may influence perceptions of law. Jurisprudential traditions influence each country's basic understandings of the concept of law, and consequently affect its view of legislative functions, the role of moral values in the legal framework, and ideas on legitimacy and the justification of legal norms. This may also influence legislative culture, even if the characteristics of these traditions in practice may not be as visible as they are in theory.

If jurisprudential traditions influence the possibilities of applying an interactive legislative approach, the question becomes which traditions create conditions that are favourable to the application of the interactive legislative approach?

---

2 C.H. Church, *The Politics and Government of Switzerland*, Ltd. Houndmills Basingstoke Hampshire, Palgrave Macmillan, 2004, pp. 55-56.

3 Church, *The Politics and Government of Switzerland*, pp. 55-56.

When analysing the legislative culture, the case-study analyzes will study each country's jurisprudential traditions as well as their legislative approaches. The examination of jurisprudential traditions mainly adopts a theoretical perspective, while the discussion of legislative approaches adopts a practical one. This dual perspective contributes to analyzing whether jurisprudential traditions indeed influence a country's legislative approaches.

4.4 The Layout

Chapter 5, 6 and 7's presentations of the case studies will be uniformly structured. First of all, each chapter will describe its country's regulation of animal biotechnology. This will involve an exploration of the legislative procedure, regulatory framework and legislative practice. The legislative practice comprises the implementation phase as well as the application of the legal norms.

Second, each chapter will discuss the regulation of animal biotechnology in light of the characterising themes of the interactive legislative approach, generally following the outline presented in Section 4.2. This will include examinations of public involvement in legislation, the role of moral values in legislation, and the legitimacy of legal norms.

Third, the analyzes will explore the political culture of the case-study countries, based on the structure presented in Section 4.3. This part starts with a description of each country's democratic traditions. It will pay particular attention to the role of the public, which will also involve an examination of the role of experts. Finally, the chapters will turn to a study of the legislative culture, which will adopt a twofold structure, discussing, first, jurisprudential traditions, and second, the country's legislative approaches in general.

Fourth and last, each chapter will present some first conclusions concerning the adequacy of the interactive legislative approach in the case-study countries. These conclusions mainly relate to the question of whether an interactive legislative approach can be recognized in these countries. However, they also involve some general remarks about the model interactive legislative approach.

5 The Regulation of Animal Biotechnology in Denmark

5.1 Introduction

This first case study explores the regulation of animal biotechnology in Denmark. Some of the characteristic tenets of an interactive legislative approach can be recognized in Danish political culture. Danish political culture is characterized by broad pluralism. Political rights such as freedom of speech and freedom of religion together with traditions of public involvement have contributed to a political culture in which all minorities can have a say. Furthermore, both citizens and experts are involved in decision-making and norm development in various ways. For example, Denmark is well known for methods of public involvement such as consensus conferences.

Chapter 4 presented the structure for analysing and discussing the case studies. This chapter will follow the plan outlined in Chapter 4. First, Section 5.2 will examine the regulation of animal biotechnology in Denmark. It will explore the legislative process, the legislative framework, and the regulation in practice. Section 5.3 will discuss public involvement in legislation, the role of moral values in legislation, and the legitimacy of legal norms. These focal points reflect the characteristic themes of the interactive legislative approach.

To understand and explain the results of the case studies' analyzes we need to understand the political context of the case-study country as well. Section 5.4 will therefore examine Denmark's democratic and legal traditions. This discussion will focus on the elements of the traditions that are relevant for understanding the regulations. As a result, it will restrict itself to assessing the prominent characteristics of the Danish legislative approach and its conformity with the interactive model of legislation. This discussion prompts an exploration of an important element of the interactive legislative approach: broad participation. As we will see, in Denmark, public participation and expert involvement are more significant in the broader political context than they are in the legislative procedure.

Finally, Section 5.5 formulates a first answer to the question of whether an interactive legislative approach can produce a qualitatively 'good' law in Denmark. These remarks will be discussed more thoroughly in Chapter 8, and in the last part of this book.

The Interactive Legislative Approach in Denmark

5.2 The Regulation of Animal Biotechnology

Animal biotechnology in Denmark is regulated in the 2005 Act on Cloning and Genetic Modification of Animals (The Danish Animal Cloning Act). This section discusses

Denmark's regulation of animal biotechnology in three sub-sections. First, Section 5.2.1 gives an overview of the legislative procedure of the Act. This overview highlights both the main reasoning in the parliamentary debate and the report of the preparatory committee that was appointed by the Minister of Science, Technology and Innovation. Second, Section 5.2.3 explores the legislation in practice, focusing mainly on the licensing procedure directed by the Animal Experiment Directorate.

5.2.1 *The Legislative Procedure*

Denmark's political agenda on animal biotechnology was first set in 1992, when the Danish Board of Technology organized a consensus conference on transgenic animals. The final report of this conference expressed some concerns about the impact of animal cloning.[1] In 1997, the birth of Dolly, the first genetically modified sheep, spurred a new round of political as well as public debate in Denmark, as it did in most European countries. The renewed debate resulted in a call by the Danish Parliament for regulation of animal cloning.[2] However, this request was never picked up. In November 2002, animal cloning was again debated in Parliament, and this debate resulted in a renewed call for regulation. Like the 1997 request, this new Parliamentary motion called on the Minister to set rules regulating the cloning of animals. Following the Parliament's request, the Danish Minister of Science, Technology and Innovation installed a preparatory committee to examine the possibilities for regulating animal cloning. Representatives of animal protection organisations, representatives of other ministries, and experts in the fields of animal ethics, animal experiments, and public health issues were appointed to the preparatory committee. It was, then, a committee founded on a notion of broad participation.

Besides evaluating the legislative possibilities for animal biotechnology, the committee was also directed to discuss animal cloning, and, more specifically, animal integrity from a moral perspective. The latter discussion was eventually supposed to lead to the development of a legal concept of animal integrity.[3] The preparatory committee finally presented the outcomes in a thorough report in 2003. This report explained the technological aspects, the relevant moral considerations, the possibilities, and the shortcomings of current regulation. Furthermore, the report included suggestions for the regulation of research in the field of animal biotechnology. This section discusses relevant outcomes of the report, which include the moral considerations expressed in the concept of animal integrity, the shortcomings of the current regulation at that time, and the advice for new regulation.

The committee's report included several interesting findings with respect to the moral considerations of regulating animal biotechnology. To begin with, the committee's report

---

1 Teknologinævnet, *Teknologiske dyr, Indgreb I kønsceller på højerestående dyr*. Slotdokument og ekspertoplæg fra konsensuskonferencen 23-25 September 1992, arrangneret af Tekologiinævnet i samarbejde med Folketingets Forskningsudvalg, København 1993.

2 *Skriftlig Fremsættelse, Forslag til lov om cloning og genmodificering af dyr*, Justitsminister, L25, 7 October 2004.

3 Ministeriet for Videnskab, Teknologi og Udvikling, *Genmodificerede og klonede dyr*, 15 October 2002, p. 4.

drew heavily on a 1996 report by the Animal Ethics Committee (AEC), an expert advisory committee established by the Danish government in 1992.[4] The committee adopted nearly all of the AEC's reasoning about the moral considerations.[5] This was not really surprising, since the chairman of the AEC was also a member of the preparatory committee, where he functioned as the sole expert on ethics.[6] The government often consults the AEC on moral issues concerning animals. The AEC has adopted a utilitarian framework for moral reasoning. In its report on animal cloning, however, animal cloning the AEC also presented a minority statement that adopted a more deontological framework.[7] Setting aside this minority opinion, the committee adopted the AEC's utilitarian majority statement.

In general, the AEC identified four moral arguments against cloning that were present in Danish public and political debates: 1) cloning simplifies nature; 2) cloning is unnatural; 3) allowing cloning will lead to a slippery slope; and 4) cloning violates the integrity of animals.[8] The Animal Ethics Committee rejected the first two arguments by comparing animal cloning with technologies that were already accepted in society and had a similar impact on nature, such as selective breeding. It disposed of the slippery slope argument with a simple decision 'not to worry about it'. Only the fourth argument about the integrity of animals was regarded as valid by the AEC members.[9] This reasoning about the four arguments against cloning was adopted by the preparatory committee and incorporated in their report.

The emphasis on integrity of animals was not surprising, as the Minister had already directed the preparatory committee to assess the moral impact of cloning in terms of animal integrity and it was therefore a pre-determined outcome.[10] The committee observed that the moral impact of animal biotechnology goes beyond the welfare and health of animals. Interference with the evolution and reproduction of animals has a greater impact on animal integrity than, for example, animal experiments. Taking into account the integrity of animals supplements concepts of health and welfare, and is therefore more suitable for expressing the moral impact of technologies such as cloning and genetic modification. Animal integrity as such may not be defined in measurable values, but nevertheless, as the committee advised, it should play a prominent role in balancing whether or not to permit a specific kind of research in which these technologies are used.[11]

Here, we can observe that the AEC's strong utilitarian approach and the Danish legislature's positivist approach are becoming 'weakly normative', as they leave room for discussions regarding fundamental values. Kemp and Rendtorff introduce this term in the context of human biotechnology, which they argue has become more open to fundamental

4 The Danish Animal Ethics Council: Information for EuroFAWC 8-9 May 2007, written by the Danish Animal Ethics Council.
5 Udtalelsse fra det dyreetiske rad, *om bioteknologi i forbindelse med dyr*, June 1996.
6 *Genmodificerede og klonede dyr.*
7 Udtalelsse fra det dyreetiske rad, *om bioteknologi i forbindelse med dyr*, June 1996.
8 *Genmodificerede og klonede dyr*, pp. 27-28.
9 *Ibid.*, pp. 33-35.
10 *Ibid.*, Ch. 2.
11 *Genmodificerede og klonede dyr*, pp. 29-35.

principles such as autonomy, integrity and dignity.[12] Despite this move toward 'weak normativity', however, the utilitarian approach continues to dominate the AEC's reasoning.[13]

The emphasis on animal integrity in the context of animal biotechnology also led to calls for special regulation to address the new and different moral impact of these technologies. These moral differences also revealed the shortcomings of existing regulation. The committee argued that due to the more extensive moral impact of animal biotechnology, these technologies should be regulated differently from other animal research. In the regulation of animal experiments, infringing on the health and welfare of animals is justified by weighing the harm to animals against the advantages for human beings and basic research. This balancing does not take into account the integrity of animals. Therefore, the preparatory committee argued for a separate regulation that would include an evaluation of this far-reaching moral impact. Arguments about risk and uncertainty supplemented the argument that animal biotechnology should be treated separately.[14]

In addition to arguing for separate regulation of animal biotechnology, the preparatory committee also made some suggestions for how the new regulation should look. Its report underlined the importance of animal integrity and the uncertainty and risks involved in cloning and genetic modification. Furthermore, the committee kept in mind the governments' policy of allowing the use of these technologies under certain conditions. Therefore, the committee argued that these technologies should be permitted, but only for a restricted set of purposes that could be justified under the principle of proportionality. The principle of proportionality can best be explained as a principle that operates by a balancing of pros and cons. It evaluates the outcome in proportion to the various interests and values at stake.

By balancing the pros and cons, and taking into account the consequences for the integrity of animals and the accompanying risks and uncertainties, the committee formulated four purposes for which the use of animal biotechnologies would be permitted: 1) basic research; 2) applied research for the improvement of human health and environment; 3) the reproduction of animals that produce material that contributes to the improvement of human health and environment; and 4) the education of people who will apply these technologies.[15] Remarkable is that despite their strong recognition during the legislative process, moral values are included here only by means of a pre-eminently political balancing of pros and cons. This would not be problematic in itself, were it not that for the fact that 'balancing the moral values' was translated here into an examination of interests. The moral impact of these technologies thus disappeared into the background. Moral values as such do not play a further role in deliberation, and seem to have been neutralized at an early stage of the legislative process.

12 J.D. Rendtorff and P. Kemp, *Basic Ethical Principles in European Bioethics and Biolaw, Vol. 1 Autonomy, Dignity, Integrity and Vulnerability*, Barcelona, Impremta Balona, 2000, pp. 161-162.

13 In human biotechnology, a movement towards a more normative approach is further developed and included in the legal framework.

14 *Genmodificerede og klonede dyr*, pp. 46-48.

15 *Genmodificerede og klonede dyr*, pp. 46-47.

The preparatory committee was appointed and directed by the Ministry of Science, Technology and Innovation. Nevertheless, it was the Ministry of Justice that finally drafted the Act on the Cloning and Genetic Modification of Animals. Officially, issues concerning animals are part of the responsibility of the Ministry of Justice, which is well known for its instrumental and pragmatic approach.[16] The Ministry of Justice followed the recommendations of the preparatory committee and included the four permissible purposes for using animal biotechnology in the Act. Furthermore, the Ministry adopted the preparatory committee's advice to create a licensing procedure. Under the Act, a license for research using cloning and genetic modification technologies could only be granted for one of the purposes listed in the legislative framework. The Animal Experiment Directorate, which was already charged with supervising the use of other animal experimentation, was appointed to make the decisions about which specific licenses would be granted.

Denmark's official legislative procedure consists of a pre-parliamentary and a parliamentary phase. In both phases the draft of a law is open for comments and questions that require a response from the Minister. This section discusses both phases together since similar concerns were expressed and similar questions were asked during each phase. Furthermore, the pre-parliamentary phase did not result in a change of draft. Additionally, the Minister's response in each phase did not differ fundamentally.

Concerns, questions, and objections to the regulation on animal cloning circled around four main issues: 1) the scope of the permissible purposes listed in the Act; 2) the lack of general principles in the legislative framework; 3) the lack of room to deliberate moral issues and risks within the licensing procedure; and 4) the responsible institution within the licensing procedure.[17] The first three issues concern the justification of the legislative framework. These issues led to debates in both the pre-parliamentary phase and in the parliamentary debate. The Minister's response to all three issues consisted of a reference to the principle of proportionality. This principle was used to justify the scope of the purposes as well as the lack of room for further debate and general principles.[18] The Minister referred explicitly to this principle as a response to the criticism. This was remarkable, as it was, after all, the inclusion of the principle of proportionality in the Act that was doubted by the critics.[19]

In addition, some members of parliament questioned whether the Animal Experiment Directorate should be the authority responsible for issuing licenses for animal cloning.[20] They doubted whether the authority responsible for overseeing animal experimentation should supervise animal cloning as well. This was especially questionable since the call for new regulation on animal cloning was spurred by the articulation of a conceptual distinction between cloning and animal experiments, particularly with respect to their moral

16 Interviews with M. Hartlev (member of the Danish Ethics Council) and with H. Anker (Professor of environmental law), Autumn 2007.

17 *Kommentaret oversigt over høringssvar om forlsag til lov om kloning og genmodificering af dyr m.v.*, Justitsminister, L8, 24 February 2005.

18 *Ibid.*

19 *Ibid.*

20 *Høringsvar vedrørende udkast til lovsforslag*, September 2004.

impacts.[21] The appointment of a new committee was suggested as an alternative. But this objection was rejected for practical reasons of efficiency. Similar expertise is needed to deal with applications for animal experiments as is required for dealing with applications for animal biotechnology. It was therefore decided that it was not necessary to appoint a different committee.

Despite this broad criticism, the Minister considered his response and especially his reference to the principle of proportionality as sufficient. The Minister did not see any need to change or complement the draft. Although the legislative procedure became weakly normative by including moral values in its deliberation, the legal texts returned to the traditional instrumental and pragmatic approach, translating moral values into interests. Following the official parliamentary procedure, the Act was accepted by Parliament on the 14th of June, 2005 and came into force in October 2005.

5.2.2 *The Legislation in Practice*

In legislative practice, Denmark's legislation on animal biotechnology includes both a phase of implementing the legal norms and a licensing procedure. In this case study, implementation and licensing run parallel. Implementation comes down to application in the licensing procedure. Therefore, this section is primarily concerned with characteristic elements of the licensing procedure. Its main focuses are the application of the four purposes for which a license can be granted, and the licensing procedure itself under the responsibility of the Animal Experiment Directorate.

The Animal Experiment Directorate consists of eleven members: a judge as chairman, five scientists from the field, and five members of animal protection organisations. The broad membership of the Directorate ensures that various viewpoints can be brought forward. The Animal Experiment Directorate is responsible for the licensing procedure for animal experiments as regulated in the Animal Experiment Act.[22]

Over the years, the Directorate has developed a standard procedure for licensing animal experiments.[23] When an application is received, the member who has the most expertise on the matter reviews the application and makes a recommendation on whether a license should be granted. Theoretically, there is room to discuss and question these recommendations, but in reality some precedents and standard reasoning have developed. Consequently, recommendations rarely lead to debates on more fundamental matters. According to the members of the Directorate, these standards have resulted in a clear framework for dealing with license applications.[24] Kornerup Hanssen explains that these standards do not involve moral deliberation, but instead include requirements for

21 *Genmodificerede og klonede dyr,* Ch. 4.

22 The Danish Animal Experiment Act.

23 Interview with A. Kornerup Hanssen (chairman of the Animal Experiment Directorate), Autumn 2007.

24 Interviews with A. Kornerup Hanssen and with H.P. Olessen (secretary of the Animal Experiment Directorate), Autumn 2007.

animals' conditions in research labs and possibilities for alternative research methods.[25] The Directorate has established a strict control mechanism to ensure that these requirements are met, and that animal welfare and health are thus protected.[26] This demonstrates again the pragmatic, instrumentalist nature of the Act,[27] which is still most prominent in questions concerning animals.

Since 2005, the Directorate has also been responsible for issuing licenses for animal biotechnology. Combining licensing on animal biotechnology with the licensing of animal experiments under one authority allowed the Animal Experiment Directorate to adapt existing standard procedures on the licensing of animal experiments to the licensing of animal biotechnology. As a result, members of the Directorate considered the implementation of the Act on Cloning and Genetic Modification of Animals fairly unproblematic.[28]

First of all, the members of the Directorate were already familiar with conducting licensing procedures. Second, they could adapt existing general experimentation standards to deal with the new animal biotechnology applications. Differences in research purposes could easily be embedded in existing procedures via an extension of the official application form.[29] Members did not see a need for further explication of the differences in moral impact between animal experimentation and biotechnology. The inclusion in the Act of four research purposes for which licenses could be granted was a signal that any moral controversies had already been dealt with. In a relatively short period of time, the Directorate established a well-functioning standard procedure for licenses in the field of animal biotechnology that has not led to any new conflicts or objections.

5.3 Characteristic Elements of a Legislative Approach

This section analyzes the three characteristic themes of the interactive legislative approach identified in Chapter 4 in light of Denmark's animal biotechnology legislation. The purpose of this analysis is to determine whether, and to what extent, Denmark employed an interactive legislative approach. It will examine, in turn, public involvement in legislation, the role of moral values in legislation, and the legitimacy of legal norms.

5.3.1 *Public Involvement in Legislation*

The preceding analysis of the Danish regulation governing animal biotechnology demonstrated that there was little public involvement in the legislative procedure and licensing procedure of the Act on the Cloning and Genetic Modification of Animals. This is a remarkable finding in a political culture that is known for its participatory methods.

25 Interview with A. Korenrup Hanssen, Autumn 2007.
26 Interviews with A. Kornerup Hanssen and with H.P. Olessen, Autumn 2007.
27 Rendtorff and Kemp, *Basic Ethical Principles in European Bioethics and Biolaw*, pp. 161-162.
28 Interviews with A. Kornerup Hanssen and with H.P. Olessen, Autumn 2007.
29 *Ibid.*

However, the public did have some indirect input. First of all, the public participated in the consensus conference on transgenic animals in 1992.[30] This meant that public opinion contributed to introducing the matter onto the political agenda. In the legislative procedure that created the Act, however, the public was represented only indirectly by the experts from different fields were consulted and involved in the development of the legal norms. Additionally, the report of the preparatory committee included the public voice by listing the public's moral concerns about animal cloning.[31]

The licensing procedure also does not set a role for the public. Again, there is a strong reliance on experts who are involved in deliberation and decision-making. The Animal Experiment Directorate is made up of experts from different fields, including proponents as well as opponents of animal biotechnology. Some of these experts are representatives of the animals' interests, but none explicitly represent the public. Even so, the public may recognize their opinions in one of the member's viewpoints.

Representation of the public is not part of the process. The experts are members of the committee, and consequently, there is no external perspective on the decision-making process. Furthermore, due to the standardization of the licensing procedure, considerations of proponents and opponents rarely lead to debates in which different viewpoints are expressed.[32]

The concept of public participation in Denmark also involves providing the public with well-organized information. In general, and also in the case of Denmark's animal biotechnology legislation, the public is informed about decision-making.[33] Most governmental documents and reports of discussion groups are easily accessible, which ensures some level of transparency in deliberation.[34] This ensures that Danish citizens are well-informed and helps to stimulate public debate. At the same time, it does not imply that the public is represented in the deliberation itself, or is truly a part of the decision-making process.

The Danish legislative approach for regulating animal biotechnology involved a relatively small, indirect role for the public. Most of the public involvement that did exist was limited to expert representation.

5.3.2 *The Role of Moral Values in Legislation*

Chapter 4 distinguished three stages of legislation in which moral values explicitly or implicitly play a role: 1) the legislative process, 2) the legal texts, and 3) the implementation

---

30 Teknologinævnet, *Tecknologiske dyr, Indgreb i kønsceller p☒ højerest☒ende dyr*, konsensuskonferencen 23-25 September 1992, rapporter 1993/1.

31 *Genmodificerede og klonede dyr,* Ch. 4.

32 Interviews with A. Kornerup Hanssen and with H.P. Olessen, Autumn 2007.

33 A. Porsborg Nielsen, J. Lassen, and P. Sandøe, 'Democracy at its Best? The Consensus Conference in a Cross-National Perspective', in *Journal of Agricultural and Environmental Ethics,* 20, 2007, pp. 31-32.

34 E.M. Basse and J. Dallberg-Larsen, 'The Danish Legal System', in H.T. Anker, B.E. Olsen, and A. Rønne (eds.), *Legal Systems and Wind Energy, A Comparative Perspective,* Alphen a/d Rijn, Kluwer Law International, 2008, p. 60.

phase. In the Act on Cloning and Genetic Modification of Animals, moral values played the most prominent and explicit role during the initial legislative process. The moral concerns raised by animal cloning were one of the incentives for designing a distinct law. The preparatory committee widely discussed and balanced moral considerations. The committee's terms of reference directed it to deliberate on the moral considerations with respect to the integrity of animals.

The amount of deliberation on moral values that actually took place, however, was limited. Both the Ministry and the preparatory committee followed the Animal Ethics Committee's recommendations on animals' integrity. Furthermore, the preparatory committee adopted the AEC's approach to moral reasoning, and used the principle of proportionality to deal with moral dilemmas.[35] Consequently, the integrity of animals was included together with other objections against the use of these technologies in a political balancing test. The preparatory committee and the Ministry adopted this policy approach for incorporating moral considerations into the legislative process and developing the list of four acceptable purposes for animal biotechnology research.

Moreover, even though animal integrity, as part of this balancing test, indirectly influenced the development of the legal norms, its moral value was translated into an interest that could be balanced in a political context. This is a remarkably utilitarian approach, which determines the morally right action by weighing the interests of all those involved. Moral values such as 'animal integrity', however, cannot be taken into account in this weighing procedure. The concept of 'animal integrity' has its origin in a different kind of normative theory, sometimes called 'deontology'. In a deontological moral theory, moral decisions are not the result of a weighing of interests. Rather, deontological theory propose moral *obligations*, for instance the obligation to respect animal integrity. Such obligations have to be honoured, and it is only permissible to deviate from them if another moral obligation takes precedence. By translating moral values, such as animal integrity, into interests that can be weighed against one another, the moral meaning of these values may be changed or downplayed.[36]

Due to the AEC chairman's prominent role, alternatives to the strong utilitarian approach were hardly discussed in the political arena. These 'other' interpretations were noted, but not included in the political and pragmatic balancing of pros and cons.[37]

Attempts to broaden the moral input in decision-making resulted in the formation of the BioTIK group, an assembly of various ethicists (including the AEC's chairman) with

35 *Genmodificerede og klonede dyr*, p. 27.

36 I owe here a special thanks to Frederike Kaldewaij for discussions about a translation of moral values into political interests and her comments on this point.

37 It must be noticed, other ethical perspectives are not ignored. As we have seen in the statement of the AEC concerning cloning in 1996 in which they also presented a minority statement. Furthermore, in the report *Genmodificerede og klonede dyr* various moral arguments are discussed. However, these arguments are easily downplayed by a utilitarian balancing from an ethical perspective (AEC) and in the political discourse by a political balancing of pros and cons. Furthermore, in academic debates there was more room to discuss various other ethical perspectives. I refer to the incentives and work of for example the Centre for Law and Ethics, Copenhagen. It seems, however, that academic debate and political and legal decision-making are different fields that do not always intertwine.

different perspectives. The BioTIK group was appointed by the government to bridge the gap between the utilitarian approach of the AEC and more deliberative approaches. In the context animal cloning, the BioTIK group suggested using the principle of proportionality. Further attempts to broaden moral input resulted in a set of ethical guidelines for decision-making. However, these guidelines did not influence decision-making about animal biotechnology.

Despite various attempts to bridge the gap between legal decision making and moral development, therefore, in the context of animal biotechnology, they are still separated discourses. Consequently, moral deliberation's influence in the legislative process was limited.

In the legal texts, moral values played a role only in the preparatory works, and were not included in the final Act.[38] The Act sets out the four acceptable purposes of research using animal biotechnology. These purposes were derived from a balancing exercise. However, the pros and cons that formed the basis for that balancing exercise were not incorporated. Instead, the moral deliberation and balancing of pros and cons that took place during the legislative procedure were considered to have settled the question of the moral impact of technologies such as cloning and genetic modification. Furthermore, the conditions for licensing included in the Act also do not include further moral deliberation.

Indeed, under the Act, moral values have no further role in licensing or implementation. Apart from a process of standardizing the conditions of the licensing procedure, there is no further development of the legal norms or reflection upon moral values within the implementation phase.[39] The licensing procedure for research using animal biotechnology, which resembles the licensing procedure for animal experiments, evaluates moral impact only in terms of health and welfare. These conditions are strictly controlled and express a concern for the wellbeing of the animals.

Downplaying moral values in the implementation phase and licensing procedure reflects a specific positivist view of law and the role of moral values in the legal context. From the perspective of the AEC chairman and the Minister of Justice, responding to new issues in the field is a matter of changing or complementing existing law, not a matter of reinterpreting and developing it.[40] Neither the chairman nor the Minister of Justice want to downplay moral values concerning animals. However, they argue that these should not be developed in the legal discourse. It is sufficient to take these moral values into consideration in the legislative procedures as part of the various interests at stake in

38 Danish Act on Cloning and Genetic Modification of Animals (2005). Preparatory works refer to the report *Genmodificerede og klonede dyr.* These preparatory reports officially belong to the legal texts and function as a source of law, see D. Tamm, 'The Danes and Their Legal Heritage', in B. Dahl, T. Melchior, and D. Tamm (eds.), *Danish Law in European Perspective*, Copenhagen, Thomson Publishers, 2002, p. 57. However, in this case the preparatory works merely function as a justification instead of a source for interpreting the legal norms in reality. In reality, only the purposes for which licensing can be granted play a role.

39 Interviews with A. Kornerup Hanssen and with H.P. Olessen, Autumn 2007.

40 Personal communication with P. Sandøe (Chairman of the Danish Council for Animals Ethics), October and November 2007.

drafting legislation.[41] A further role for these moral values is unnecessary in the Danish perspective.

An analysis of various academic debates and reports in the fields of plant and food biotechnology and animal cloning reflects that the moral impact of animal biotechnology is in fact widely acknowledged and discussed among Danish academics and practitioners.[42] Nevertheless, the Danish pragmatic approach still seems to dominate the legislative practice, especially within the Ministry of Justice. Regulation does not involve a normative framework, and moral values as such are not included in regulation. Neither do these values play a role in implementation.

5.3.3 *Legitimacy of Legal Norms*

It is possible to identify three sources for legitimating legal norms in the Danish context. The first one requires the following of official legislative procedures as a precondition for ensuring the validity of legal norms.[43] This relates to the dimension of legal validity identified in Chapter 2's framework on the quality of law. The second source for legitimacy is derived from broad deliberation. This source of legitimacy is not explicitly acknowledged by the Danish legislature. However, broad deliberation brings about both a popular legitimacy and a substantive justification, which explains the Danish emphasis on this aspect of lawmaking.[44] Third, in regulating animal biotechnology, the legislature uses a principle of proportionality to justify legal norms substantively, which characterizes the approach of the Ministry of Justice.[45] This section discusses the latter two sources, since these are most important for legitimating the legal norms concerning animal biotechnology.

Broad deliberation contributes to legitimating legal norms in two ways. First, it establishes popular legitimacy. Popular legitimacy here is not established by the classical reading of direct public involvement in terms of acceptance or approval. Instead, it is embodied in the process of norm development through a broad deliberative process in which the public can recognize their viewpoints.[46] In the context of Denmark's animal biotechnology legislation, the preparatory committee anticipated the viewpoints of the public, although these were not further included in the legal context. Additionally, this

41 Personal communication with P. Sandøe, October and November 2007.

42 For example, see reports of the Danish Centre for Bioethics and Risk Assessment, C. Gamborg, M. Gjerris, J. Gunning, M. Hartlev, G. Meyer, P. Sandøe, G. Tveit, *Regulating Farm Animal Cloning: Recommendation of the Project,* Danish Centre for Bioethics and Risk Assessment, 2006; G. Tveit, K.H. Madsen, P. Sandøe, *Biotechnology and the Public Debate*, Danish Centre for Bioethics and Risk-Assessment, 2003.

43 Tamm, 'The Danes and Their Legal Heritage', p. 56.

44 Basse and Dallberg-Larsen, 'The Danish Legal System', p. 60.

45 Interview with. H. Anker, October 2007. It is not only used in approaching animal biotechnology, but a principle that is generally acknowledged in Danish administrative law. See Basse and Dallberg-Larsen, 'The Danish Legal System', p. 63.

46 Porsborg Nielsen, et al., 'Democracy at its Best?', pp. 22-23; 29.

broad process and its emphasis on transparency provide insights into the legislature's reasoning, which also improves popular legitimacy.

Second, broad deliberation can establish substantive justification. According to deliberative democratic traditions, broad deliberation ensures a well-considered decision.[47] Input from various fields of expertise guarantees that decisions are thought through, and are, therefore, justified substantively from various points of view.

The most significant justification of legal norms in Denmark is the principle of proportionality. In the context of animal biotechnology, this principle was broadly accepted and applied in several contexts. First of all, the Animal Ethics Committee has applied this principle to justify its utilitarian analysis when advising on moral issues. Second, in the legislative context, the Minister of Justice used this principle to balance pros and cons to decide whether a certain norm could be justified. The Ministry of Justice seemed to use this principle from more pragmatic motives, out of a preference for clear norms.[48] And third, the Minister referred to this principle in parliamentary debate when responding to criticism against the draft Act. The principle of proportionality justifies the four purposes for which animal biotechnology can be employed. Additionally, the principle is applied to justify the absence of room for further deliberation within the licensing procedure on fundamental matters concerning the status of animals.[49]

The principle of proportionality sets the content of legal norms, and therefore provides substantive justification. In contrast with substantive justification as explicated in Chapter 2's framework, the proportionality principle lacks content. The framework explains substantive justification in terms of reflection in light of critical morality or in light of policies. Both must be understood as methods for substantiating decisions. The content of the policies or the content of the critical morality justifies decisions or norms substantively. Proportionality is a more procedural principle, and is in itself empty of fundamental substantive grounds. In Denmark, the principle of proportionality mainly seems to take on a political character derived from pragmatic motives. It is a political method for dealing with controversial issues. It does not favour one value or a set of values, but instead weighs them against each other.[50]

This understanding of the principle of proportionality contrasts with the ideal-type of the interactive legislative approach, in which substantive values play a more prominent

---

47 J. Cohen, 'Deliberation and Democratic Legitimacy', in J. Bohman and W. Rehg (eds.), *Deliberative Democracy, Essays on Reason and Politics*, Cambridge Mass. The MIT Press, 1997.

48 The Minister of Justice and the AEC use the principle of proportionality for different reasons: pragmatic motives resp. adopting a utilitarian point of view. Utilitarian and pragmatic arguments are two different things, although in this chapter ' pragmatic', 'utilitarian' and also ' instrumentalism' seems to be mixed up. I am aware that utilitarian arguments are not always most pragmatic. However, concerning the principle of proportionality in the Danish context these arguments strongly intertwine. In some papers concerning the normativity of Danish law, these two types of arguments are mentioned within the same context, see for example Rendtorff and Kemp, *Basic Ethical Principles in European Bioethics and Biolaw*. To adjust to the Danish terminology, this chapter mentions these types of arguments within the same context as well.

49 *Justitsminiteriets kommenterede oversigt over høringssvar; Høringsvar vedrørende udkast til lovsforslag*, September 2004.

50 *Genmodificerede og klonede dyr*, Ch. 4.

role in justifying legal norms. However, it must be noticed that, in general, Danish bioethics and biolaw is becoming more open to fundamental principles as a point of departure for legal reasoning.[51] In human biotechnology and health law, various fundamental principles, such as respect for personal autonomy and integrity, structure decision-making. In health law, these principles form the basis of formal procedures that emphasize the importance of informed consent.[52]

5.4 DANISH POLITICAL CULTURE

This section discusses Denmark's political culture, democratic traditions and legislative culture. It also explores the role of public involvement in the broader Danish political context. This involves the role played by experts, since expert consultation and public involvement are correlated in Danish politics. This elaboration may contribute to explaining the paradox that while Denmark is well known for its participatory legislative methods, at the same time, the public play a limited role in regulating animal biotechnology.

5.4.1 *Democratic Traditions*

Denmark is often characterized as a deliberative democracy. This section presents the arguments supporting this claim. These arguments focus mainly on theoretical foundations, but additionally offer some concrete examples to illustrate these theoretical statements.

Officially Denmark is organised as a representative democracy. Similar to most western countries, Danish supreme power is based on a Trias Politica. Three independent organs counterbalance one another: the executive, the legislative, and the judicial powers.[53] The Danish Parliament, the *Folketing*, has the legislative power and is the sole organ empowered to legislate. However, in reality, the Danish executive can initiate legislation as well.[54] The *Folketing*'s 179 parliamentarians are elected by the Danish people based upon proportional representation. This system guarantees that all Danish regions as well as representatives of any large, organized minority groups obtain seats according to the number of votes cast. With this system the Danes acknowledge the principle that it is up to the majority to make decisions. However, there is also a possibility to call for referenda, which gives the Danish people the opportunity to directly influence decision-making.[55] In reality, referenda are hardly ever called for, but the potential functions as a control

51 Rendtorff and Kemp, *Basic Ethical Principles in European Bioethics and Biolaw*, pp. 161-162.

52 M. Hartlev and U. Lind, 'Use of blood samples from the Danish PKU-Biobank. A Study of the Conceptualization of Research in Law and in Action', Paper for a workshop of EURECA, Tenerife, November 2005, pp. 1-28.

53 Basse and Dallberg-Larsen, 'The Danish Legal System', pp. 75-78.

54 <www.folketinget.dk>, The Danish Acts as enacted by the Folketing only take effect after receiving the Royal Assent. In reality, the Monarch is beyond the three powers, but Her Majesty The Queen officially exerts authority e.g. when appointing or dismissing the Ministers of Danish government.

55 Basse and Dallberg-Larsen, 'The Danish Legal System', p. 76.

mechanism. This possibility for control has led to a trusting relationship between Danish citizens and their elected representatives. Furthermore, elected representatives as well as governmental institutions are easily accessible, and provide relevant information to the public.[56]

Andersen et al. argue that the representative role of the *Folketing* has weakened in recent years. On the one hand, this results from the growing unification of the European Union, which has diminished the *Folketing*'s power from above. On the other hand, it results from the decentralisation of competence to local authorities, which has weakened the representative role of the *Folketing* from below. In addition, active membership in political parties has declined, and this in turn has changed the perception of representation in Parliament. Political parties are no longer viewed as representing their grassroots support, but rather the desires of single voters.[57] At the same time, public participation on the local level has become more prominent.[58]

Andersen et al. also argue that we can recognize deliberative decision-making processes in Danish politics. They base their argument on March and Olsen's distinction between integrative and aggregative political processes.[59] Aggregative processes can best be explained as processes that distil the will of the people through bargaining between rational citizens who reason from a self-interested perspective.[60] In integrative processes, on the other hand, the will of the people is discovered through deliberation rather than bargaining. The people's perspective does not involve self-interest, but instead shared social values.[61]

Representative democracies are built on aggregative political processes with decisions based on majority rule. However, it is integrative political processes that can be observed most prominently within Danish democracy. March and Olsen relate these processes to the ideals of participatory democracies. In deliberative democracies, a specific kind of participatory democracy, these integrative political processes play a central role. Integrative political processes identify the interests and preferences of the people that may influence decision-making. Such deliberative processes characterize Danish democracy.[62] Citizens have various possibilities for participating in all kinds of discussions, especially on the local level.

Nielsen et al., who studied Denmark's consensus conference on genetically modified organisms, also argue that Danish political processes echo the values and principles of deliberative democracies.[63] One such deliberative ideal is based on the idea that priority should be given to processes that stimulate better arguments. Open communicative

56 <www.folketinget.dk>.

57 I. Andersen and B. Jaeger, 'Scenario Workshops and Consensus Conferences: Towards more Democratic Decision-making', in *Science and Public Policy*, 26/5, October 1999, p. 333.

58 Andersen and Jaeger, 'Scenario Workshops and Consensus Conferences', p. 333.

59 J.G. March and J.P. Olsen, *Rediscovering Institutions: The Organizational Basis for Politics*, New York, The Free Press, 1989.

60 Andersen and Jaeger, 'Scenario Workshops and Consensus Conferences', p. 333.

61 March and Olsen, *Rediscovering Institutions: The Organizational Basis for Politics*.

62 Andersen and Jaeger, 'Scenario Workshops and Consensus Conferences', p 333.

63 Porsborg Nielsen Porsborg Nielsen, et al., 'Democracy at its Best?'.

and deliberative processes exemplify these kinds of processes. Nielsen et al. state that in Denmark, the legitimacy of decision-making stems from the idea that deliberative processes improve the quality of decision-making. They argue that the involvement of lay people may not necessarily lead to the inclusion of the voice of the people in legislation, but nevertheless stimulates deliberative processes and debates on political matters in the public arena.[64] Deliberation is considered a way of allowing the better argument to come forth, and creates possibilities for all voices to be heard. Participating in political debate is the Danish way to 'do politics', which can ensure that citizens become active citizens.[65] In Denmark, these methods seem to work well and engender positive attitudes towards political decision-making.

The Danish Board of Technology has developed several other participatory methods in the context of technology assessment that stimulate integrative political processes. The *Folketing* set up the Board of Technology in 1995 to replace the previous board (*Teknologinaevnet*) formed in 1986. This Board, which is monitored by the Ministry of Science, Technology and Innovation, advises the Ministry and Parliament on legislative and policy matters concerning science and technology. Furthermore, the Board aims to build a bridge between experts and lay people using all kinds of participatory methods, such as the consensus conferences, citizens' panels, and scenario-workshops.[66] These methods are tools for institutionalising public involvement and interweaving debate and deliberation with citizens into the political process. These integrative processes are an important feature of Danish political culture, and reflect Denmark's democratic ideals.

5.4.2 *The Role of the Public*

In general, the public does not play any official role in the legislative process, the enactment of the law, or the licensing procedure in Denmark. Exceptions are the provisions in the Constitution for referenda on constitutional changes, though these are seldom called for.[67] Nevertheless, the public's lack of official status in decision-making does not imply that public opinion has no value or that Danish respect for the public is a matter of lip-service. Rather, the bulk of public involvement comes from deliberative participation. In Danish political culture, the deliberative participation of the public helps to ensure popular legitimacy by increasing popular support.[68] This section explores the role of the public and its relation to deliberative notions of public involvement.

The Danish ideology of public participation finds its roots in the principle of 'people's enlightenment', or '*folkeoplysning*'. This principle is based on the assumption that a

64 *Ibid.*, p. 31.
65 *Ibid.*, p. 29.
66 Interview with L. Klüver (Director of the Danish Board of Technology), October 2007.
67 Det Danske Selskab, *Nordic Democracy, Ideas, Issues, and Institutions in Politics, Economy, Education, Social and Cultural Affairs of Denmark, Finland, Iceland, Norway and Sweden*, Copenhagen, 1981, p. 127.
68 L. Klüver, et al., EUROPTA. *European Participatory Technology Assessment*, The Danish Board of Technology, 2000.

democracy can only function well where the population consists of well-educated and engaged citizens.[69] Besides influencing the political framework, this principle impacts the daily life of the Danes. Public education, participation, and debate are intertwined in the social context as well. This ideology has led to a culture of discussing everything. This has also influenced the perception of the public's role. Just being part of debate is already a means of establishing public involvement, which is an important source for public support.[70] Giving the public an official say in the decision-making process is not required for gaining public support.

According to Fearon, deliberation involves a particular kind of discussion in which all pros and cons are weighed carefully against one another.[71] In deliberative processes, several individuals discuss an issue together and aim to come together to well-considered, public-spirited decisions. The public can have a role in this deliberation, directly or indirectly, but does not need to have a decisive say. Indeed, none of the actors involved in these deliberative processes seem to have a decisive say. This reasoning seems to apply to the Danish interpretation of public involvement as well.

The deliberative notion of public involvement in Danish politics is most prominently reflected in elements such as inclusiveness and transparency, which guide Danish deliberative processes.[72] Inclusiveness does not necessarily imply that *all* viewpoints and arguments are reflected in decision-making.[73] Neither does any individual voice need to have a decisive say. The Danish idea of inclusiveness involves discussing everything in a broad context. Inclusiveness entails equal possibilities for all actors to express their opinions, their concerns and post their statements. All relevant viewpoints and arguments – including those of the public – are considered and weighed, but do not necessarily have to influence decision-making.

The transparency of deliberative processes has also become an important element of public involvement. Transparency brings insights into the reasoning of the decision maker. Additionally, transparency ensures that all parties have access to all relevant information for decision-making. For the Danes, transparent dialogue is one of the hallmarks of active democracy.[74] It is so important that transparency in terms of having access to all relevant information has been translated into a right to information.[75] Expert groups are often appointed to advise the government on policy matters, and to inform the public as well. For example, the Danish Council for Ethics has an important role in informing

69 Det Danske Selskab, *Nordic Democracy*, p. 127.

70 *Ibid.*

71 J.D. Fearon, 'Deliberation as Discussion', in J. Elster (ed.), *Deliberative Democracy*, Cambridge, Cambridge University Press, 1998, p. 63.

72 O. Renn, T. Webler, H. Rakel, P. Dienel, and B. Johnson, 'Public Participation in Decision Making: A Three-Step Procedure', in *Policy Sciences*, 26, 1993, pp. 189-214; E. Einsiedel, 'Citizen Voices: Public Participation on Biotechnology', in *Notizie di Politeia*, 2001, pp. 1-15.

73 *Ibid.*, pp. 189-214.

74 A. Porsborg Nielsen, J. Lassen, and P. Sandøe, 'Involving the Public. Participatory Methods and Democratic Ideals', <bioethics.kvl.dk>, p. 7.

75 Basse and Dallberg-Larsen, 'The Danish Legal System', p. 60.

the public. Providing comprehensive information stimulates public debate, which in turn provides insights into the concerns and questions of the public.[76] Public debate as such can eventually feed into decision-making.

The Danish people evaluate their role positively. Danes feel they are taken seriously, and consequently feel involved.[77] Furthermore, the public can recognize their statements and concerns in legislative deliberation, and therefore feel that they are represented in decision-making processes. Therefore, citizens are more willing to comply with the rules.

However, we could question whether this description of public involvement adequately reflects reality. In reality, inclusiveness and transparency have not given the public opportunities to participate in the decision-making process comparable to those of experts. Experts are often appointed to participate on committees that advise the government on all kinds of policy matters, while the public has no such role. For example, the consensus conference about transgenic animals may have set the political agenda, but did not influence law-making substantively.[78]

Consensus Conferences

The organisation of consensus conferences reflects the perception of the role of the public in Danish political culture. Internationally, consensus conferences are a Danish showpiece. Danish consensus conferences consist of three phases.[79] In the first phase, the preparatory phase, a panel of lay people is informed about the issue at hand and is asked to prepare a list of questions for additional information. The second phase is twofold: first a panel of experts addresses the questions, and second, they present their answers to the lay panel at a public conference. In the last phase, the lay panel withdraws and discusses the answers of the experts. Finally, the lay panel presents its outcomes. Outcomes that are based on consensus are published in an official document on the consensus conference. Deliberations on features that did not result in a consensus are not included in the final document. This official document is distributed among governmental institutions and all kinds of public organisations.[80] Therefore, the impact of consensus conferences is not limited to public debate, but can also have significant influence on policy-making, for example by setting the political agenda.[81]

The first purpose of consensus conferences is related to public support. Involving the public in decision-making establishes a positive attitude towards decisions, even when the involvement does not affect the decision-making. Consensus conferences are said to open

76 Terms of references of the AEC. Chapter 5, The Danish Act on the Protection of Animals, 1992.

77 Porsborg Nielsen et al, 'Democracy at its Best?', pp. 30-31. See also S. Joss and J. Durant, Introduction, in S. Joss and J. Durant (eds.), *Public Participation in Science. The Role of Consensus Conferences in Europe*, London, Science Museum, 1995 on the symbolic value of participatory methods such as consensus conferences.

78 Teknologinævnet, *Tecknologiske dyr, Indgreb i kønsceller p⊠ højerest⊠ende dyr.*

79 L. Klüver, 'Consensus Conferences in the Danish Board of Technology', in S. Joss and J. Durant (eds.), *Public Participation in Science. The Role of Consensus Conferences in Europe*, London, Science Museum, 1995.

80 *Ibid.*

81 *Ibid.*

up possibilities for citizens to be involved in decisions that concern them. Furthermore, these conferences may help to improve public awareness of the complexity of a policy problem and its controversies.

Second, consensus conferences are seen as a method for improving deliberation. Deliberation is valuable in itself, because it is considered a method for allowing the best argument to come forth. Furthermore, public involvement in deliberation may overcome decision-makers' lack of experiential knowledge, which citizens can provide. Klüver relates these two purposes to an instrumental and intrinsic value of consensus conferences.[82] Furthermore, Joss distinguishes a third value of consensus conferences: a symbolic value. Methods such as consensus conferences instil in the public a feeling of being involved in decision-making. Citizens feel their concerns and considerations are seriously taken into account.[83]

Einsiedel argues that methods such as consensus conferences can meet deliberative democratic ideals.[84] In deliberative democratic models, public reasoning and debate among equal individuals establish the legitimacy of decisions. In her argument, Einsiedel emphasizes that deliberative processes are a kind of learning process.[85] The consensus conference in particular reflects elements of a learning process.

Joss even argues that the Danish model of consensus conferences goes beyond the meaning of 'public participation'. In general, he reasons, public involvement is translated in terms of providing information. Public participation is a method for increasing access to information, which is needed for decision-making. According to Joss, consensus conferences create further possibilities for citizen involvement. In a consensus conference, citizens can be actually involved in as well as actually influence decision-making.[86]

However, Joss's interpretation of consensus conferences and specifically their impact on public participation is questionable. From a theoretical point of view, he may be right. The aims of consensus conferences and their ideals seem to point to a level of public participation in which the public can influence decision-making. In reality, however, their influence is limited. To start, the extent of their influence is doubtful. For example, the consensus conference on transgenic animals may have set the Danish agenda on animal biotechnology, but the considerations expressed during the consensus conference were neither embedded in the reasoning of the preparatory committee in 2003 nor in the political debate drafting the Act.[87]

Furthermore, even if the outcomes of the consensus conference had been included in the regulation, they still would not adequately represent the public's opinion. The official

---

82 *Ibid.*

83 Joss and Durant, 'Introduction', pp. 14-15.

84 E. Einsiedel and D. Eastlick, 'Consensus Conferences as Deliberative Democracy: A Communications Perspective', in *Science Communication,* 21/4, June 2000, pp. 326-327.

85 E. Einsiedel, 'Publics at the Technology Table: Consensus Conference in Demark, Canada and Australia', in *Public Understanding of Science*, 10, 2001, pp. 83-84.

86 S. Joss, 'Public Participation in Science and Technology Policy- and Decision-Making. Empheral Phenomenon or Lasting Change?', in *Science and Public Policy,* 1999, pp. 290-292.

87 Teknologinævnet, *Tecknologiske dyr, Indgreb i kønsceller på højerestående dyr*; *Kloning*, Udtalerser fra Det Etisk Råd of Det Dyreetiske Råd (2000); *Genmodificerede og klonede dyr.*

documents contain only those points on which the conference actually came to a consensus. Issues that led to major or even small controversies and even those on which a large majority exists are not included.[88] The considerations in the final report are therefore limited, and cannot adequately represent public opinion. These restrictions make the inclusiveness as well as transparency of these consensus conferences doubtful.

The Role of Experts in Decision-Making

Traditions of deliberative democracy have strengthened the experts' role in Danish decision-making. The involvement of experts can contribute to a transparent dialogue and the inclusion of various viewpoints. Moreover, it ensures a horizontal decision-making process. This section distinguishes at least two roles for experts in the Danish political context: 1) advising the government on policy matters and in decision-making; and 2) stimulating public debate and informing the public.

The 'advising' role of experts is most explicit. Experts from different fields are often appointed to advise the government on policy and legislative matters.[89] Their statements can directly influence decision-making; however, governmental institutions cannot be forced to adopt experts' recommendations. Advisory committees exist in various forms. For example, a minister of government can appoint a preparatory committee to give advice on a specific legislative matter. This type of committee is often temporary and may have clear directives to guide its task. Members come from various backgrounds: experts from the field, representatives of all kinds of public organisations, and representatives of governmental institutions. Such a committee was formed to advise the Danish government on the regulation of animal cloning.

Furthermore, committees can be appointed to advise the government on certain fields of policy. These committees have a more indefinite character. They are often independent, but are monitored by the Ministry responsible for their field of policy.[90] Most often, these experts are consulted for their expert knowledge on part of a policy problem. Consequently, they come from more or less similar backgrounds, and address only part of the issue at stake.

The Danish Animal Ethics Committee (AEC) represents such a committee, and is often consulted regarding the ethical considerations concerning animals. Its terms of reference and assembly are regulated in the fifth chapter of the Animal Protection Act, which directs the AEC to advise the Ministry of Justice on the ethical considerations involved in matters concerning the protection of farm animals, pets and laboratory animals. The AEC presents comprehensive statements including overviews of current legislation. These statements discuss all relevant moral considerations. The AEC provides insights into all kinds of ethical viewpoints including concerns that may exist among the public. The committee's aim is to present a unanimous and consensus-based advisory position reflecting the most rational viewpoint.

---

88 Klüver, 'Consensus Conferences in the Danish Board of Technology'.

89 Andersen and Jaeger, 'Scenario Workshops and Consensus Conferences', pp. 331-334.

90 Such as the Danish Animal Ethics Councils and the Danish Council of Ethics.

In general, the AEC adopts a utilitarian approach. As explained in Section 2.1, the AEC employs the principle of proportionality to deal with both the negative consequences and the benefits of the use of animal cloning technologies. The AEC's approach closely resembles the instrumentalist and pragmatic approach of the Minister of Justice, which enhances the influence of the committee's recommendations on the Minister's policy. On the one hand, this resemblance to the approach of the Minister of Justice ensures that moral concerns are at least taken into account. On the other hand, the only moral concerns that are taken into account are those presented by the AEC.[91]

Although the AEC's members are appointed as experts on animal ethics, other viewpoints about ethics do exist. The Danish government aimed to broaden the ethical input by appointing the before-mentioned interdisciplinary expert group BioTIK. This group's main task concerned the development of general ethical guidelines, and resulted in the publication of the report 'an ethical foundation for genetic engineering choices'.[92] Furthermore, BioTIK provided advice about animal cloning.[93] This expert group consisted of members of several expert committees, councils, and research centres in the field of biotechnology and/or ethics, including members of the AEC. The mixture of experts from different groups was supposed to assemble an expert group in which all viewpoints were represented. It was expected that this cooperation could address the controversies between viewpoints on the moral dilemmas posed by genetic engineering, and would result in the production of ethical guidelines that could satisfy all actors.

BioTIK's report had an ambiguous influence on Danish policy. The utilitarian approach of the AEC and the instrumentalist approach of the Danish government were resisted by some of BioTIK's experts.[94] Despite the fact that they were supposed to bridge the gap between utilitarian and deliberative thinking, the expert group rejected the utilitarian approach. It presented a discussion paper with five guidelines for ethical decision-making in the field of genetic engineering.[95] These prestigious guidelines referred to the importance of democratic debate and specifically public debate, the status of animals, and openness of decision-making. However, according to some of the members of this interdisciplinary expert group, the gap between viewpoints of the various experts was too large. As a result, BioTIK's guidelines were more or less vague and empty concepts that none of the

91 The AEC is not in itself a player in the academic debates, however its members are. For example, the chairman is also connected to the Danish Centre for Bioethics and Risk Assessment; the Centre adapts to academic debates as well as it plays a role in advising the government on bioethical issues and stimulating public debate. See note 41.

92 Its report '*An Ethical Foundation for Genetic Engineering Choices*', was published by the Ministry of Trade and Industry, Denmark, 1999.

93 BioTIK expert group, *De genteknologische valg*, et debatoplæg udarbejedet af BioTIK-gruppen, Ehrversministeriet (Ministry of Environment), June 1999.

94 Especially in the academic research centres the rather pragmatic approach was rejected. Some of these academics were also involved in committees and councils in a political context. Interviews with M. Hartlev, J. Rendtorff, P. Kemp, P. Sandøe, and G. Meyer, Autumn 2007.

95 *An Ethical Foundation for Genetic Engineering Choices,* 1999.

members was really convinced of.[96] BioTIK emphasized that the guidelines were the result of political action to stimulate co-operation.[97] According to them, the outcomes were, in fact, a political compromise instead of a substantive set of guidelines. In reality, the influence of these guidelines was limited. For example, the Act on Cloning and Genetic Modification of Animals does not reflect the guidelines. Instead, the utilitarian approach of the AEC was used to address moral concerns in the field of animal ethics.

At the same time, the Ministry of Trade and Industry has published a Statement on Ethics and Genetic Engineering in which the Minister adopted the outcomes of BioTIK.[98] This Statement was debated by Parliament in April 2000, and received broad general support among the political parliamentary parties. Furthermore, Kemp, who was also a member of BioTIK, together with Rendtorff underlined the importance of these ethical guidelines in introducing concepts such as dignity and vulnerability into the general biotechnology debate.[99] Kemp and Rendtorff argue that BioTIK's guidelines established that the 'pragmatic bureaucratic legal system has slowly become more open to ethical issues'.[100] They acknowledge that Danish legislators still have a tendency toward a more utilitarian approach for addressing ethical issues. The basic ethical principles, though, have become a more 'integrative part of bioethics and biolaw'.[101]

The difference in influence of the AEC, on the one hand, and BioTIK, on the other, reflects the dual structure of Danish politics. For concrete regulatory matters, the Danish Ministry of Justice favours an instrumental, pragmatic approach in which concrete and clear advice is preferred. Law and morality are considered two different fields. Although the moral impact of law is acknowledged, legal discourse is not considered the right context for developing moral norms. A utilitarian approach to the moral issues, therefore, seems more adequate for incorporating moral concerns in the instrumentalist, pragmatic approach of the Danish Ministry of Justice.

At the same time, decisions need to be well considered. Deliberation is an important pillar of Danish democracy, and is valued as an end in itself. Broad participation is therefore required as well, which involves input from a variety of perspectives. Broad participation is more evident in the context of general biolaw than it is in the regulation concerning issues with animals.[102] Decision-making in other Ministries, such as 'Trade and Industry' and 'Interior and Health' is slowly becoming 'weakly normative'.[103]

96 Interviews with participants of the BioTIK Expert Group, P. Sandøe and G. Meyer, Autumn 2007.

97 Interviews with participants of the BioTIK Expert Group, P. Sandøe and G. Meyer, Autumn 2007.

98 The Danish Ministry of Trade and Industry, 'The Danish Government Statement on Ethics and Genetic Engineering', August 2000.

99 Rendtorff and Kemp, *Basic Ethical Principles in European Bioethics and Biolaw*, p. 169.

100 *Ibid.*, p. 170.

101 *Ibid.*, p. 170; T. Achen, *Den bioetiske udfordring. Et retspolitik studie af forholdet mellem etik, politik og ret I det lovforberedende arbejde vedrørende bio- og genteknologi i Danmark, Norge og Sverige*, Linköping, 1997, p. 103.

102 Rendtorff and Kemp, *Basic Ethical Principles in European Bioethics and Biolaw*, pp. 160-170.

103 In these Ministries, genetic engineering in general is organized and regulated. The Danish Ethics Council is also under responsibility of the Ministry of Interior and Health, which uses a more open approach.

With regard to the second role of expert committees, the Danish government appoints committees to broaden deliberation with an explicit focus on stimulating public debate and informing the public. A first example is the Danish Board of Technology, discussed above in Section 5.4.1, which is directed to develop tools that can bridge the gap between scientists, the public and politicians.

An example of an expert committee that combines both roles and focuses on both public debate and advising the government is the before-mentioned Danish Council of Ethics.[104] The Council of Ethics was established in 1988 and its terms of reference involved providing the Danish Parliament, official authorities, and the public at large with ongoing advice and information about the ethical problems raised by developments in the National Health Service and the field of biomedicine. The Council of Ethics is monitored by the Minister of Health and Interior, but the Minister has no instructional powers. In general, the Council functions independently. However, a parliamentary committee appointed to safeguard relations between the Council of Ethics and the Parliament is empowered to appoint a certain number of Council members and can also add topics to their terms of reference.

The Council of Ethics aims to provide insights into the various ethical viewpoints that may exist on a topic.[105] Therefore it has deliberately chosen not to present clear and unanimous advisory statements. Instead, the Council attempts to stimulate both public and political debate by providing a stocktaking of various arguments and viewpoints. Here, we can recognize a different approach than is issued by the AEC. It is therefore an interesting example of the Danish decision-making system becoming more open to moral deliberation in the political.

5.4.3 *The Legislative Culture*

A country's legislative culture constitutes its perceptions of law and its legislative approaches, both of which influence its ideas on the legitimacy of legal norms and the functions of law. Legislative culture is influenced by the interplay between democratic and legal traditions. Section 5.4.1 discussed Denmark's democratic traditions. This section will explore Danish legal traditions, and further explain the pragmatic legislative approach that is prominent in Denmark.

Legal systems are traditionally divided into two legal families: common law and civil law. Arguments for classifying the Danish legal system into either of these traditions can be made, and it must be admitted that continental legal thinking has had a considerable influence on the Danish system. However, the most common opinion today seems to point towards a third legal family: Nordic law.[106] It is rather difficult to point out how Nordic law

---

104 <www.etiskraad.dk>, interviews with members of the Danish Council of Ethics, M. Hartlev and U. Hybel, Autumn 2007; see also The Act on the Danish Council of Ethics.

105 Interviews with M. Hartlev and with U. Hybel, Autumn 2007.

106 Tamm, 'The Danes and Their Legal Heritage', pp. 42-43.

differs from other legal families. However, two prominent characteristics can be stressed. First of all, Nordic legal systems do not emphasize legal formalities, and sources of law other than the traditional case law, constitutional law, and regulatory law play a prominent role. And second, the lack of general parts in the civil law codes of the Nordic countries created possibilities for a more pragmatic way of thinking.[107]

Danish legal traditions are generally characterized as being positivistic and pragmatic.[108] The ideas of legal philosopher Alf Ross had a heavy influence on Danish legal science and additionally the conception of law. His ideas were in turn influenced by American legal realism, and led to a highly positivist and at the same time pragmatic view about law. Ross' approach was important for the development of Scandinavian legal realism. His legal theory became influential in the twentieth century and has resulted in a pragmatic, down-to-earth, and realistic conception of law.[109]

This resulted in a conception of law as a concrete set of norms that is explained in terms of effectiveness of the norms, and its validity depends on probabilities for future application of the norms.[110] By adopting a pragmatic conception of law, norm development begins with an appreciation of the issue at stake. Furthermore, the positivist view is recognized in the potency of norms, which is supposed to increase if norms are based on a well-established doctrine or on a 'statutory provision whose interpretation has been established in long and consistent practice'.[111]

According to Ross, morality has no institutional character and is therefore a completely different concept. For example, he refused to recognize medical ethics as a source of law.[112] Morality is an individual phenomenon that lacks power to exercise force and does not embody norms of competence. There may be a considerable degree of harmony between the concepts of law and morality since both are rooted in the tradition of the community's culture. Furthermore, both concepts stand in a relationship of reciprocal co-operation. Nevertheless, Ross argues that law and morality do not correspond. First of all, the 'purpose' of law is to create a community and to keep its peace. Morality, on the other hand, is a personal phenomenon and can therefore easily cause conflicts in society. Thus, morality runs contrary to community. Second, law has the tendency to rationalize the use of norms and legal concepts. Morality desires concrete adjustment to particular circumstances and is, consequently, more open for change.[113] However, recent developments have shown that, especially in the field of biolaw, norm development has become more open and normative.[114]

---

107 *Ibid.*, p. 43.

108 Rendtorff and Kemp, *Basic Ethical Principles in European Bioethics and Biolaw*, pp. 160-170.

109 Basse and Dallberg-Larsen, 'The Danish Legal System', p. 65.

110 A. Ross, *On Law and Justice*, The Lawbook Exchange, Ltd., Clark New Jersey, 2004 (reprint), Ch. 2.

111 *Ibid.*, p. 45.

112 A. Ross, 'Livets Hellighed contra individets autonomi', in *Ugeskrift for læger*, Copenhagen, 1979.

113 Ross, *On Law and Justice*, pp. 59-64.

114 Rendtorff and Kemp, *Basic Ethical Principles in European Bioethics and Biolaw*, pp. 160-170; *An Ethical Foundation for Genetic Engineering Choices*.

The influence of Scandinavian legal realism on current legislative practice has resulted in a concept of law which presents law as concrete norms focused on effectiveness. Symbolic values are less prominent in this concept of law. Norm development starts in the field. Morality can be part of this field, but is considered different from legal norm development.

Since society is often confronted with complex moral issues, however, ethics has become integrated into legal norm development as well.[115] Decision-making processes on issues such as human biotechnology and health are already more open to including moral reflection within the regulatory framework. For example, in health law principles such as privacy, autonomy, and integrity are important values that play a role in developing the regulatory framework as well as its field of application. Although the Danish legislature still often uses a pragmatic approach in regulating these issues,[116] it has become more open to moral considerations. For example, in a case study about the legal framework for using tissue samples, Hartlev and Lind conclude that decisions about whether blood samples may be used in research depends less on the legal distinction between research and non-research uses, and more on questions of moral acceptability.[117] This may not amount to the inclusion of moral values in the legal framework, but it shows that moral acceptability does play a role in decision-making.

In sum, the Danish legislative approach is mainly instrumentalist. Both Denmark's legislative procedure and its conception of law reflect an instrumentalist and pragmatic approach. Legislation is closely connected to reality and consists of a concrete set of rules that can easily be applied in practice.[118] In general, legislation is used as an instrument in political reform work and does not endeavour to state any fundamental values or shared customs.[119] At the same time, recent developments have established that legal norm development is slowly becoming weakly normative, as it includes moral values in deliberations about regulation.

Both deliberative democratic traditions and Scandinavian legal realism have influenced the relationship between legislative and norm development processes in Denmark. As was made clear in the analysis of the Act on Cloning and Genetic Modification of Animals, norm development ends with the enactment of legal norms. In general, norm development is limited to broad deliberation during the legislative procedure. In these legislative processes, the influence of legal realism can be observed. The involvement of experts and the public reflects a strong interaction with the field. This interaction increases the effectiveness of legal norms, and, consequently, their validity.[120]

---

115 Achen, *Den bioetiske udfordring*, p. 103.

116 J. Rothmar Hermann and M. Rowlandson, 'The Role of Ethics and Morality in EU Law', in *Journal of International Biotechnology Law*, 5, 2008, p. 245.

117 Hartlev and Lind, 'Use of Blood Samples from the Danish PKU-Biobank. A Study of the Conceptualization of Research in Law and in Action'.

118 Basse and Dallberg-Larsen, 'The Danish Legal System', p. 65.

119 *Ibid.*

120 Ross, *On Law and Justice*, pp. 34-35.

Interaction within the Danish legislative approach is, however, limited to the legislative process. Deliberation in the Danish legislative process does not involve interaction in the ongoing process of norm development. In Ross's terms, future application does not imply future developments of norms.[121]

5.5 Conclusion

This chapter has discussed Danish regulation of animal biotechnology, focusing on its interactive elements and its relationship with the general Danish political culture. This analysis indicates that Danish legislative approach for regulating animal biotechnology does not correspond with the ideals of the interactive legislative approach. A partial explanation can be found in the Minister of Justice's pragmatic and positivist view of law.

The Danish Minister of Justice's approach to law differs in several ways from the ideal-typical model of the interactive legislative approach. These differences concern the extent of the norm development procedure, the use of open norms in the legal context, and the role of experts. In general, Danish legislation acknowledges an instrumentalist function of legislation. In this approach, the enactment of legal norms entails the closure of norm development. This runs counter to the basic premise of dynamics, which is one of the two normative ideals of the interactive legislative approach. In the ideal-typical interactive model, open norms are incorporated into the legal framework to increase possibilities for further discussion and concretisation of legal norms. Norms have therefore both an expressive and communicative function. In Denmark, the instrumental function of law and the pragmatic legislative approach do not support or leave room for open norms. Instead, concrete rules are preferred. Additionally, the Danish perception of law influences the role of moral values in legislation as well as ideas on the legitimacy of legal norms. Moral value plays a prominent role in the legislative process. However, their role is part of a balancing of pros and cons.

This analysis of the regulation of animal biotechnology may colour the general impression of Danish legislative approaches. We have seen that in general legal decision-making is more open to moral issues and is slowly becoming 'weakly normative'. Furthermore, the Danes are known for their participatory methods that ensure broad deliberation and popular legitimacy. Several characteristics of the interactive legislative approach can be recognized in Danish political culture. For example, interactive elements include an emphasis on deliberation in legislative processes as well as in Danish politics. Denmark uses several methods of involving and informing the public, which satisfies the public and engenders a positive attitude of the citizens towards legal norms. Decision-making in Denmark is not solely dominated by the legislature or governmental bodies, and includes consultations with various experts.

121 Ross, *On Law and Justice*, p. 45.

At the same time, Danish political culture is still primarily pragmatic, and this pragmatism takes precedence over broad deliberation. For example, attempts to broaden the debate on animal biotechnology by including diverse ethical viewpoints got bogged down in a political compromise between experts. Because of this, the inclusion of these different viewpoints has not made a significant impact on decision-making concerning animal biotechnology. The gaps between the experts were too large to be bridged. Expert involvement may be broad, but does not guarantee that expert input is reflected in the eventual decision.

While traditions of deliberative democracy deliberative democracy may provide the appropriate conditions for the interactive element of an interactive legislative approach, they do not seem to allow sufficient space for the dynamic element to function adequately. The dynamic element requires a more normative legislative approach not so tied to pragmatism and instrumental reason.

6 The Regulation of Animal Biotechnology in Switzerland

6.1 Introduction

At first sight, Switzerland seems to fit some of the characteristics of the ideal-typical model of the interactive legislative approach. Switzerland is pre-eminently a country that is characterized by broad (segmented) pluralism. Swiss culture is characterized by the disparities between German-speaking, French-speaking, and Italian-speaking Swiss. These disparities have resulted in a country that is both linguistically and politically divided. Elements such as executive power-sharing in broad coalitions, federalism, and direct democracy have resulted in a political system in which the various cultural groups can be represented in decision-making, and in which their cultures can be preserved. The autonomy of the 26 Cantons exemplifies this system most explicitly. A mixture of direct democracy and federalism has led to a consensus model of decision-making that guarantees a high level of public participation. Referenda and popular initiatives are powerful instruments of direct public involvement. This system ensures that all viewpoints – including those of traditional minorities – are taken into account in decision-making.

This case study explores the Swiss regulation of animal biotechnology: the Gene Technology Law (GTL). Similar to Chapter 5's exploration of Denmark's animal biotechnology regulation, this chapter will be structured by the plan outlined in Chapter 4. First of all, Section 6.2 introduces the GTL. It examines its legislative process, its regulatory framework, and its functioning in practice. Second, Section 6.3 analyzes the role of the public in legislation, the role of moral values, and the Swiss concept of legislative legitimacy. Section 6.4 explores Swiss political culture in order to provide a better understanding of the outcomes of the previous sections. This exploration assesses the most prominent characteristics of the Swiss approach, and analyzes whether and to what extent they are in conformity with the interactive model. Finally, Section 6.5 will discuss some preliminary outcomes of this case study. These outcomes will be discussed more thoroughly in Chapter 8, and in Part III of this research.

The Interactive Legislative Approach in Switzerland

6.2 The Regulation of Animal Biotechnology

The Swiss regulation of animal biotechnology, the Gene Technology Law (GTL), was influenced by three major political events: a popular referendum on a constitutional amendment in 1992, a popular referendum on a constitutional amendment in 1998, and the

'Gen-Lex' motion launched in 1998.[1] The first constitutional referendum required the regulation of genetic engineering to ensure the safety of humans, animals, and the environment, as well as the protection of animal and plant species. This amendment was adopted, and resulted in Article 120 BV (Art. 24 old Swiss Constitution), under which regulation of genetic engineering has to take into account the 'dignity of living beings' (*Würde der Kreatur*).

The second constitutional amendment was an attempt at responding to the 1992 call for regulation on gene technology. The amendment was rejected in June 1998. This amendment, which was initiated by the Schweizerische Arbeitsgruppe Gentechnik (SAG),[2] proposed a general ban on genetic engineering. The SAG argued that respect for the dignity of living beings could only be assured through a general ban on this technology.[3] At the same time, a majority of Swiss people feared that such a ban would hinder medical research. Moreover, they feared it would result in an economic setback for pharmaceutical companies, which would be forced to move entire research programs abroad.

At the same time, the Parliament adopted the Gen-Lex motion, which presented a package of several amendments to laws relevant for genetic engineering. By presenting an alternative, the Parliament's motion probably influenced the rejection of the second constitutional amendment.[4] It was this motion that finally resulted in the enactment and adoption of the Gene Technology Law (GTL) in March 2003.

This section discusses the GTL. First of all, Section 6.2.1 explores the legislative procedure through which the regulation was adopted and the legal norms as put down in the legal framework of the GTL. Second, Section 6.2.2 describes the legislation as it functions in the licensing procedure. This exploration mainly focuses on the licensing procedure as directed by the Federal Office for the Environment (FOEN) of the Department of Environment, Transport, Energy and Communications (DETEC).

6.2.1 The Legislative Procedure

As soon as the Gen-Lex motion was accepted in 1998, the Swiss Minister of Environment, Transport, Energy and Communications (the Minister) presented a draft regulation

---

1 F.X. Perrez, C. Errass, and K. Bende, 'GMO-regulation: Case Study of Switzerland', December 2003, <www.law.nyu.edu>, p. 3.

2 Switzerland has several working groups similar to the SAG that are independent organizations that function as a critical platform for certain political themes. These platforms initiate several actions for political change, but also provide information to the public. The SAG is focused on subjects circling around gene technology. The Working Group reflects a critical attitude towards gene technology and often argues for gentech free surroundings.

3 D. Ammann and A.F. Goetschel, SAG-Studienpapiere, *Die Verfassungsnorm der Würde der Kreatur/ Konsequenzen für die Zulassung genmanipulierter Tiere,* 1998.

4 In reality, a popular iniative is often responded to by a motion of the parliament. An explanation for this is found in the quality of the popular iniatives. As soon as Parliament doubts whether an iniative can lead to a success in terms of effective regulation it aims to come with an alternative that come forward to the concerns that are expressed by the people (by iniating a draft), but is at the same time a more effective and well-considered one.

to several parties for pre-parliamentary consultations (*Vernehmlassungsverfahren*). The Gen-Lex motion attempted to provide a comprehensive framework for the use of GMOs in both science and industry. This framework aimed to guarantee the protection of human health, animals and the environment; respect for the dignity of living beings; the protection and sustainable use of biodiversity; and the provision of factual and transparent information to the public.[5] The draft regulation on biotechnology included a list of Gen-Lex's general aims. Furthermore, it proposed various additions to existing regulations to make them more responsive in the context of biotechnology. The draft focused on labelling and certification requirements, liability rules, access to information, the approval of GMOs for food and agriculture, and the substantive concretisation of the constitutional provision on the 'dignity of living beings'.

The pre-parliamentary consultation procedure resulted in 176 reactions by political parties, public organisations, the Cantons, the Federal Courts, private organisations, and even a few private persons. In general, most of these actors accepted the draft. But it also caused some controversies. To begin with, some reactions called for a distinct regulation on gene technology.[6] Some expressed doubts about the liability sections of the draft regulation. And some questioned the extent of the protection of human beings, animals, the environment, and biodiversity, as well as the meaning of 'dignity of living beings'. Concerning the latter, some suggested that there was a need for further explication of 'dignity of living beings', while others suggested leaving the interpretation up to expert committees.

The Minister also consulted the Ethics Committee on Gene Technology in the Non-Human Field (ECNH) about the moral impact of the proposed regulation. The ECNH, which would ultimately have a prominent role in defining and deciding on moral controversies and concerns in licensing, is a committee consisting of about twelve members from different fields of expertise. The members are appointed to give advice on the moral impact of issues concerning non-human gene technology.

At the end of the pre-parliamentary procedure, the ECNH presented a statement on the concept of the dignity of living beings. The statement presented dignity of living beings as a gradual concept with human dignity on top. Next came the dignity of 'higher' animals, which closely resembled the interpretation of human dignity, then the dignity of 'lower' animals and plants, which the ECNH interpreted as protection of the species and could better be explained in terms of value.[7]

Despite the seeming unanimity of the report, however, the committee members actually had not yet come to an agreement on defining and clarifying the concept of the

5 The Gene Technology Law is directed to gene technology in the non-human field and mainly refers to GMOs. GMOs in the legal context mean 'organisms in which the genetic material has been altered in a way that does not occur under natural conditions by crossing or natural recombination'. With the terminology organisms is referred to animals, plants and micro-organisms. Perrez, Errass, and Bende, 'GMO-regulation: Case Study of Switzerland', p. 17; Art. 5 GTL.

6 Eidgenössisches Volkswirtschaftsdepartement, 'Bericht über die Ergebnisse des Vernehmlassungsverfahrens zum Vorentwurf der Gen-Lex-Vorlage', Bern, 1998.

7 ECNH, Vorläufige Stellingsnahme zur Gen-Lex-Vorlage, September 1998.

dignity of living beings. Consequently, not all members supported this interpretation.[8] The ECNH was aware of its limits and the controversial nature of its work, and therefore emphasized that further definitional work was still required. To bridge the lack of consensus, the ECNH suggested the development of a set of criteria – for example, harm to integrity or natural behaviour – that would assist in evaluating whether the dignity of living beings had been violated in an individual case. Furthermore, the ECNH formulated a set of purposes for which a violation of the dignity of living beings can be justified: medical research, biological research, improvement of for the lives of people in developing countries, ecological advantages, improvement of the quality of food, and gene-pharming.[9]

The Minister ultimately incorporated most of the outcomes of the pre-parliamentary procedure (with the exception of the suggestion that biotechnology should be regulated under a distinct law) in the draft non-human gene technology regulation. The pre-parliamentary procedure can therefore be characterized as an interactive process with broad involvement. The suggestions of the participants of the pre-parliamentary procedure were all taken seriously and considered during the formulation of the final draft.

The modified proposal was presented to Parliament in 2000, and spurred a long and controversial debate that lasted for the next three years. First, the proposal was submitted to the Committee for Science, Education and Culture of the Council of States, one of the two chambers of the Swiss Parliament.[10] After eighteen days of deliberation, the parliamentary Committee came up with a draft regulation of gene technology in the non-human field. The Committee argued that only a distinct regulation could fill the gaps of the existing Environment Protection Act (*Umweltschutzgesetz*). It also included some major and controversial issues such as introducing a moratorium on the use of biotechnology; increasing the protection of human health, animals and the environment; including distinct provisions on the dignity of living beings, rights to information, and liability rules; and a step-by-step approach to licensing.[11] With the exception of the moratorium, the Council of States approved all of the Committee's changes.

8 There are different translations of the concept in French and English, which already point out the difficulties with the interpretations of this concept. I choose to adopt to the translation living beings to avoid concepts with a theological veil.

9 ECNH, Vorläufige Stellingsnahme zur Gen-Lex-Vorlage, September 1998.

10 The Swiss Parliament is made up of two chambers: the Council of States and the National Council. The Council of States consists of 46 members. Twenty cantons have two representatives and the 6 half-cantons each elect one representative. The National Council consists of 200 members who are directly elected by the people. In the official legislative procedure, amendments and changes made by the Council Committees go back and forth between the Councils until agreement is reached. The order of deliberation in both the Councils is interchangeable. This last phase in the Swiss parliamentary debate is the process of *Differenzbereinigung*, in which the remaining differences between the Councils are resolved. A law can only be enacted if both Councils adopt the draft, and therefore, both Councils have to agree. Each Council can discuss a proposal three times. If the Councils then still disagree, an *Einigungskonferenz* is organized, in which 13 members of both Councils come together and work on a compromise on which both Councils have to vote.

11 The latter adopts to the approach as already used in the Environmental Protection Act. Concerning the Gene Technology Law, the step-by-step approach builds on the idea that GMOs are firstly used in a contained system, secondly in experimental field tests, and are finally marketed. Every step has to be successfully concluded before taking the next step and, consequently, require different authorizations. Safety requirements increase for each step. Perrez, Errass and Bende, 'GMO-regulation: Case Study of Switzerland', p. 25.

The draft was then sent on to the National Council, the other chamber of the Swiss Parliament. The National Council's Committee for Science, Education and Culture deliberated on and further refined the draft, which was approved by the National Council and sent back to the Council of States. For the next few years, the proposal continued to be sent back and forth between the Councils and Committees.

The parliamentary debate on the GTL centred on five themes: the moratorium; requirements for the approval of experimental field-testing and use of GMOs in agriculture; the protection of GMO-free production from contamination by GMOs; liability rules; and the right of NGOs to appeal the approval of GMOs intended for use in the environment.[12] The Councils could also not agree on the purposes of the law. They disputed whether the GTL should be a law for supporting research or, instead, for protecting human health, animals, the environment and biodiversity. Finally, the Councils found a compromise between these purposes, and in March 2003 both Councils accepted the draft. The GTL entered into force on the first of January 2004.

The Gene Technology Law sets up a comprehensive legal framework that includes extended sections on the purposes, licensing procedure, and terminology of the law.[13] According to Errass, the Swiss legislature adopted a process-based approach that restricted the scope of the law and ensured a kind of control via a licensing procedure.[14] This process-based approach used general principles and procedural provisions to ensure that the law was open to further interpretation. Furthermore, the procedural and open character is reflected in the GTL's enforcement provisions, its appointment of two expert committees for consultation, and its step-by-step approach.[15]

The use of a process-based approach was remarkable, as it was not in line with the traditional Swiss product-based legislative approach. Due to the complexity of the issue, the Minister was aware of the need for further interpretation of some norms as well as for procedural decision-making techniques such as licensing. At the same time, it seems that the Minister still preferred the traditional product-based approach.[16] The Minister tried to provide a concrete set of rules for avoiding interpretative conflicts. To begin with, the GTL explicates three purposes for which genetically modified vertebrates may be produced: research, therapy, and diagnostics on or for humans and animals. These criteria

12 In the end, the moratorium was cut out as well as the right for NGOs to appeal.

13 GTL, 1-01-2004.

14 C. Errass, *Offentliches Recht der Gentechnologie im Ausserhumanbereich,* Bern, Stämpfli Verlag AG Bern, 2006, p. 114.

15 First of all, general principles of both protection of various elements and respect of dignity of living beings are ensured within the framework of licensing. Second, the process-based approach can be recognized in provisions in which the liability, the licensing-procedure, its enforcement, and its judicial context are regulated. Third, provisions on enforcing the legal norms include several duties for providing information as well as the appointment of two expert-committees that should be consulted in the licensing procedure: the ECNH and the Federal Expert Committee on Biosafety (SECB). And fourth, provisions on the procedure of licensing following a step-by-step approach reflect a process-based approach.

16 Errass, *Offentliches Recht der Gentechnologie im Ausserhumanbereich,* p. 114. As Errass argues, the product-based approach is considered the ideal-type, while the process-based is the 'real-typisch' solution for dealing with issues such as animal biotechnology.

were fleshed out with provisions concretising the extent and meaning of protection of humans, animals, the environment, and biodiversity; the protection of GMO-free production; and freedom of choice. The dignity of living beings was further defined as referring to both animals and plants, but in different ways. These provisions complemented the list of purposes for which gene technology may be used with: 1) a set of concrete criteria that need to be established to ensure biosafety; and 2) a set of interests for which infringements of the respect for dignity of living beings can be justified.

As a result, the regulation consists of a remarkable combination of detailed procedural regulations, concrete norms on the purposes of the law, and norms that reflect general fundamental principles.[17]

6.2.2 *The Legislation in Practice*

In the Swiss legislative practice, the norms contained in the Gene Technology Law are implemented through a licensing procedure. Implementation and licensing, thus, run parallel.

The Biotechnology Section of the Federal Office for the Environment is the authority responsible for licensing.[18] The Biotechnology Section has to take into account three main points in deciding on licensing applications: expert opinion; financial coverage of precautionary measures by means of which any hazards or impairment can be identified, averted, or eliminated; and the provision of information to the public. In the licensing procedure, the first two points play an essential role.[19] First, the Biotechnology Section verifies whether the applicant can financially cover the necessary precautionary measures to assure biosafety. Then, it consults various federal offices that may be involved in order to seek their opinion on the consequences of the application to their field of responsibility, such as agriculture or economics. Furthermore, the Canton in which the research will take place can express its concerns about or suggestions for licensing. Here too, we can recognize an interactive process of broad involvement.

Most prominent and important in the licensing procedure is the consultation of the two expert-committees.[20] The consultation process starts with the Swiss Federal Expert Committee for Biosafety (SECB), which advises the Biotechnology Section concerning biosafety risks. The SECB must be consulted as part of each license application since biosafety is one of the conditions that must be ensured before granting a license. Consequently, the SCEB has a strong influence on licensing.

After the biosafety consultation, the ECNH is consulted on moral matters. The ECNH is not obliged to give advice on each license application, and therefore restricts itself to

---

17 *Ibid.*, p. 130.

18 Comments on the licensing procedure by Hans Hochbach (director of the Biotechnology Section), interview Spring 2007.

19 Interview with Hans Hochbach, Spring 2007.

20 Both these Committees are seated at the same office as the Biotechnology Section, which makes consultation easy accessible.

controversial or new issues raised by particular applications.[21] As a result, the ECNH can respond to new trends and can think through the consequences and moral impact of these trends. At the same time, the ECNH is responsible for the moral side of the licensing applications. In effect, this means that a moral evaluation is only performed if the ECNH classifies an application as morally controversial. Consequently, if the ECNH refrains from giving advice, the Biotechnology Section can presume that the license application is not controversial and does not require further moral deliberation.

Finally, the Biotechnology Section decides whether or not to grant a license by taking into account the outcomes of these consultations. Licensing authority is thus both formally and in practice allocated to the federal office. And the work of the expert committees takes place outside this setting. As a result, the expert committees are not restricted to deliberating within a legal context, and can also deliberate within their own fields of expertise.[22]

6.3 Characteristic Elements of a Legislative Approach

This section examines the Swiss legislative approach to animal biotechnology by exploring the Gene Technology Law in the context of the three characteristic elements of the interactive legislative approach: public involvement in legislation, the role of moral values, and legitimacy of legal norms.

6.3.1 *Public Involvement in Legislation*

Switzerland's long tradition of direct democracy has led to the development of several popular rights such as referenda and popular iniatives. Popular approval is one of the foundations of Swiss regulation and decision-making. However, these popular rights are primarily directed at providing legitimacy for the legal system, and do not necessarily guarantee the Swiss people direct influence on decision-making. In context of the Gene Technology Law, the public did play a direct role in the legislative process, but it was a restricted one. At the same time, the public played a prominent indirect role, as Swiss citizens were represented in the process via various organisations that were involved in the pre-parliamentary procedure.

As discussed in the previous sections, the GTL resulted from a popular constitutional amendment calling for regulation of gene technology in the non-human field. By approving this amendment, the Swiss expressed their desire that respect for the dignity of living beings should be included in legislation governing the use of reproductive and genetic material from animals, plants and other organisms. The Committee for Science, Education and Culture of the National Council took the Swiss people's concerns into account

21 Interview with A. Willemsen (secretary of the ECNH), January 2007, March 2007; Bovenkerk and Poort, 'The Role of Ethics Committees in Public Debate', pp. 29-30.

22 Bovenkerk and Poort, 'The Role of Ethics Committees in Public Debate', pp. 28-30.

when initiating the Gen-Lex motion. This approach ensured public support for the motion. In other words, taking into account the people's concerns in drafting the law avoided a subsequent call for a referendum. Although the public was not officially involved in the legislative procedure that created the GTL, the Swiss people indirectly consented to the regulation. The people could recognize their concerns and preferences in the draft GTL. And if they had not, they could have called for a referendum.

The public's only official role[23] in licensing is their right to information.[24] Applications for licenses as well as the outcomes of consultations with the expert committees must be made public. Both expert committees and the responsible federal office are directed to inform the public of relevant information and to stimulate public debate. Consequently, the people have ample opportunity to become fully informed. The expert committees provide extensive information to the public and ensure that the information is easy accessible. However, they do not actively stimulate public debate. As a result, public input in decision-making is lacking.

Other organizations do, however, act to stimulate public debate. For example, Gen-Suisse – a private organisation that is partially independent, though financially supported by pharmaceutical companies and research institutions – aims to bridge the gap between scientists and the public.[25] The forum provided by Gen-Suisse is not be equivalent to an official public role in decision-making. Nevertheless, the work of Gen-Suisse and other organisations ensures a vivid public debate.

How does the level of public involvement in Switzerland relate to public involvement as hypothesized in the ideal-typical model of the interactive legislative approach? In the legislative process that created the GTL and the licensing procedure, the public's official role was limited. In the legislative process the public played almost no role, despite their ability to call on several official popular rights. Overall, therefore, direct public involvement was restricted, and public opinion was primarily represented through indirect channels.

This lack of involvement can be partially explained by the strong Swiss tradition of establishing public support. Since public support was already established by the public adoption of the 1992 constitutional amendment calling for a regulation on gene technology, further public involvement was not required.

Another explanation relates to the two conceptions of 'the public' (collective and fragmented) distinguished in Chapter 4. The Swiss seem to conceive of 'the public' as one actor, to whom various popular rights are attributed. The role of the public as such legitimates the Swiss political system. At the same time, the Swiss also acknowledge the conception of 'the public' as a mixture of various actors who can be represented in all kinds of organisations. These representative groups influenced the pre-parliamentary procedure.[26]

---

23 By an official role I understand the possibilities to express concerns or being consulted in licensing as has been prescribed in the regulatory framework.

24 Art. 18 GTL.

25 See <www.gensuisse.ch/>, Gen-Suisse for example organizes every year Tage der Genforschung. Furthermore, they organize research days during which the visit research institutes are open for the general public. The days are organized to make biotechnology better accessible.

26 See Church, *The Politics and Government of Switzerland*, p. 56.

Under the conception of a single public, public involvement in Switzerland mainly seems to take the form of public support and approval, rather than public input. This leads to a perception that public involvement in regulating animal biotechnology was restricted. At the same time, however, a relatively large percentage of Swiss citizens join public and private organisations. The local fishing club can and actually does express their concerns in the political arena.[27] The legislative process and licensing procedure set out in the GTL reflect an interactive and horizontal process in which a large number of actors were involved (176 responses in the pre-parliamentary procedure). Under the second conception of 'the public', therefore, citizens were actually involved in the legislative process, and had a prominent role in designing the regulatory framework.

6.3.2 *The Role of Moral Values in Legislation*

Chapter 4 distinguished three stages of legislation in which moral values explicitly or implicitly play a role: 1) the legislative process, 2) the legal texts, and 3) the implementation phase. In Switzerland moral values played a role in the regulation of non-human gene technology in all stages.

During the GTL's legislative process, moral values played an explicit role. This started with the constitutional provision on the dignity of living beings and the call for further regulation. As discussed in Section 6.2.1, because it was one of the starting-points for regulation, the dignity of living beings was intensively discussed during the pre-parliamentary procedure. However, it was not further developed in the official legislative procedure. The discussions that took place during the pre-parliamentary procedure were seen as sufficient for addressing the moral impact of animal biotechnology and did not lead to further debate.

The dignity of living beings also plays a prominent role in the legal texts. 'Respect for the dignity of living beings' is presented as one of the purposes of the GTL. Furthermore, Article 8 GTL concretizes the 'respect for the dignity of living beings' as well as its scope. According to that Article, the dignity of living beings is harmed if species-specific traits, functions, or habits are substantively impaired, and that impairment is not justified by a legitimate interest.[28]

Article 23 GTL also ensure a prominent role for moral value, as it appoints the ECNH to advise the Federal Council on enacting regulations, and the federal and cantonal authorities on enforcement from a moral point of view. Furthermore, the ECNH is directed to engage in public dialogue about the moral issues of biotechnology.[29] These provisions

---

27 As it actually did during the pre-parliamentary procedure concerning the GTL, see Eidgenössisches Volkswirtschaftsdepartement, 'Bericht über die Ergebnisse des Vernehmlassungsverfahrens zum Vorentwurf der Gen-Lex-Vorlage', Bern, 1998.

28 In Art. 8 GTL the legitimate interests are put down; 1) human and animal health; 2) the securing of sufficient food; 3) the reduction of ecological impairment; 4) the conservation and improvement of ecological living conditions; 5) a substantial benefit to society at the economic, social or ecological level; and 6) the increase of knowledge.

29 Art. 23 GTL.

ensure a moral reflection on the use of gene technology, which goes beyond the legal context. As previously pointed out, the ECNH is not limited to commenting on licensing applications, but can examine current regulation in light of valid rationales outside the context of legislation, which aligns with the idea of justice.[30]

These provisions also ensure a prominent role for moral values during the implementation phase. Officially, the concept of 'respect for dignity of living beings' and the ECNH's terms of reference leave room for further norm development. And indeed, exploration and development have not ended with the ECNH's statement in the pre-parliamentary procedure. Various standards concerning licensing have been developed over time. For example, genetic modification of mice for medical research on cancer is no longer considered morally controversial.[31] Furthermore, the ECNH and various associated committees have published several documents defining and exploring the dignity of animals.[32]

Concerning the presence of moral values, therefore, the Swiss case study seems to correspond with the ideal-typical model of the interactive legislative approach. However, some caveats related to the role of moral values must be made. To start, the development of moral standards has not necessarily involved the justification and concretisation of the gradual concept of dignity, but has instead been mainly restricted to application in practical cases. Norm development concerning the dignity of living beings has been limited to discussions focusing on practical cases. In the licensing procedure, both the ECNH and the authorities have adopted the ECNH's pre-parliamentary interpretation of dignity, which was said to require further development. The ECNH's interpretation has therefore functioned as a framework for further interpretation during the licensing procedure. This has resulted in the early stagnation of norm development concerning the dignity of living beings, despite the fact that the meaning of dignity as a gradual concept was still controversial and has not yet been clarified. The lack of a clear meaning is evident in, for example, the variety of translations of this concept in English (dignity of creature, creation, living beings) and French documents (l'integrité de la créature, dignité de la créature).

It must be mentioned that the ECNH has published a few statements presenting a more general reflection on the dignity of living beings.[33] Furthermore, associated academics have contributed to defining and exploring the concept.[34] However, these explorations have mostly been confined to academic debate on the extent of the gradual concept as it applies in particular cases, and have not addressed the fundamental substance of the concept itself. As a result of the ECNH's initial advice, the substance of the 'dignity of living

---

30 C. Errass, 'Zum Verhältnis von Recht und Ethik in der Verfassungsbestimmung über die Gentechnologie im Ausserhumanerbereich', in *ZSR* 2002/I, p. 339.

31 Interview with K. Buerki (member of the ECNH), April 2007.

32 For example see a joint statement of the ECNH and the Swiss Committee on Animal Experiments, *Dignity of Animals,* 2001, and; *Research on Primates–An Ethical Evaluation,* 2006.

33 *Dignity of Animals,* 2001, and; *Research on Primates–An Ethical Evaluation,* 2006.

34 For example, P. Balzer, K.P. Rippe, P. Schaber, *Menschenwürde vs Würde der Kreatur: Begriffsbestimmung, Gentechnik, Ethikkommissionen,* Freiburg/München, Alber Verlag, 1998; P. Balzer, K.P. Rippe, and P. Schaber, 'Two Concepts of Dignity for Humans and Non-Human Organisms in the Context of Genetic Engineering', in *Journal of Agricultural and Environmental Ethics,* 13, 2000. pp. 7-27.

beings' was set as meaning a gradual progression from lower to higher dignity for lower and higher forms of life. Despite the fact that the ECNH had not come to a consensus on this definition, and, indeed, emphasized the need for further development of the concept, the only thing left open to question after the pre-parliamentary phase was the extent of dignity protected. This was problematic for two reasons. First, input in the (pre-parliamentary) debate was limited. The experts who contributed to the debate constituted a limited group of players in the academic field, and represented only one set of viewpoints. One of these players was chairman of the ECNH and thus also represented the Committee's viewpoint. Second, the fact that the ECNH presented their statements as being based on a consensus among the members[35] contributed to the foreclosure of further debate. The strong drive for consensus precluded attempts to alter the definition and the essence of the 'dignity of living beings'.

A good example is the ECNH's statement about a license application for a field trial with primates by Zürich's ETH in 2007.[36] Both the ECNH's own discussion and its response reflected fundamental disagreements on the interpretation of the definition of 'dignity of living beings'. It was not clear at all whether the interpretation of 'dignity of living beings' as a gradual concept was justified or how it could be explained in the individual case. This shows that the ECNH's interpretation was not based on consensus among the parties: instead, parties were diametrically opposed.[37] The strong drive for consensus resulted in constructing 'dignity of living beings' as a gradual concept as if it were based on a consensus. In reality, this consensus was superficial, and many divergences still existed. The search for consensus prematurely stifled further exploration of the concept, and undermined reflection on the concept's moral meaning.[38]

In sum, it seems at first sight that moral values play a prominent role in all legislative stages in Switzerland. It also seems that there is enough room for moral deliberation and further norm development in and beyond the legal context to support an interactive legislative approach. However, further analysis of the GTL's implementation process points to other results. In reality, there is little ongoing reflection on or interpretation of moral concepts.

6.3.3 *Legitimacy of Legal Norms*

In Switzerland, the legitimacy of legal norms is generally explained in terms of popular legitimacy and legal validity. In addition, although less obviously, substantive justifications are used to justify Swiss legal norms.

---

35 Interviews with A. Willemsen and with K.P. Rippe (chairman of the ECNH), Spring 2007.

36 Interview with A. Willemsen, March 2007.

37 Some of the members of ECNH partly disagree, but nevertheless, the statements of the ECNH are presented to the public and the authorities as a unanimous decision, as has became clear during personal communication with members of the ECNH, 2007.

38 Bovenkerk and Poort, 'The Role of Ethics Committees in Public Debate', p. 27.

The Swiss political system is built on its people. The people are the source of legitimacy in decision-making.[39] At the same time, following official procedures is an important feature of the Swiss system for validating legal norms. Consequently, popular legitimacy is incorporated in Swiss official legislative procedures. The Constitution is the basis for the Swiss legal system. Thus, official procedures that ensure popular and cantonal rights in decision-making are put down in the Constitution.[40]

Switzerland's strong orientation towards popular legitimacy results from its mixture of federalism and direct democracy.[41] This has led to a preference for processes of negotiation and compromise to resolve conflicts.[42] In a nutshell,[43] the emphasis on popular rights and federalism contributed to the development of the Swiss consensus model, in which decision-making is based on deliberation among the major political forces, eventually leading to consensus. To minimize the risk of a referendum, the decision-making process is open to all kinds of groups.[44] This has resulted in popular legitimacy and consensus becoming leading concepts for justifying legal decisions. This is true for legitimacy in general, as well as in the GTL's legislative process.

Popular legitimacy played an important role in legitimating the GTL's legal framework. The GTL's pre-parliamentary procedure resulted in a broad range of statements by various parties. The legislature attempted to resolve controversies between viewpoints by an open deliberative process that was oriented toward achieving a final consensus. This process reflected a high level of integration of various viewpoints about non-human gene technology.

Substantive justifications were also used to legitimate the GTL's legal norms. The GTL's legal framework includes a list of fundamental values that need to be guaranteed. These values reflect the purposes of the law and act as guidelines for further decision-making. Consequently, they provide substantive justification of the tenets as well as the scope of the GTL.

A few remarks need to be made relating to the role of moral values in providing substantive justification for the GTL. First of all, the substantive justification of the GTL was not derived only from moral reflection. It was also oriented toward policies.[45] Values such as the dignity of living beings, despite earlier moral reflections, became part of a balancing of pros and cons. The Swiss legislature took a balance-of-interests approach to controlling the use of gene technology instead of banning it altogether.[46] Despite the SAG's argument

39 Church, *The Politics and Government of Switzerland*, p. 55.

40 *Ibid.*, p. 143.

41 See Section 6.4.3.

42 J. Reich, 'The Interactionist Model of Direct Democracy. Lessons from the Swiss Experiences', SSRN, 2008, p. 16, p. 31.

43 In Section 6.4 this development will be discussed more thoroughly. For now, I ask to reader to follow my claims.

44 W. Linder, 'Direct Democracy', in U. Klöti, P. Knoepfel, H. Kriesi, W. Linder, Y. Papadopoulos, P. Sciarini (eds.), *Handbook of Swiss Politics*, Zürich, Neue Zürcher Zeitung, 2007, p. 113.

45 See for an explanation of the various dimensions on which the quality of law may depend, Chapter 2 of this book.

46 Eidgenössisches Volkswirtschaftsdepartement, 'Bericht über die Ergebnisse des Vernehmlassungsverfahrens zum Vorentwurf der Gen-Lex-Vorlage', Bern, 1998, p. 31.

that adopting dignity as an intrinsic value should lead to a general ban, as any research with animals could not be justified, the legislature declined to incorporate the moratorium in the draft GTL.[47]

Second, moral values were included in the legislative framework in order to ensure popular legitimacy. These values reflect a variety of concerns expressed by the people represented in organized groups. Including these values in the GTL was designed to ensure public support, rather than provide substantive justification. At the same time, including the concerns expressed by the people reveals the expressive function of law. This expressive function is not directly a substantive justification, but it is related to the content of the norms.

In sum, the GTL was primarily justified by means of popular legitimacy. Though substantive justification also played a role, it was more limited due to the ambivalent role of moral reflection in the legislative procedure and implementation process.

6.4 Swiss Political Culture

This section discusses Swiss political culture. To start, Section 6.4.1 analyzes the Swiss form of government. Section 6.4.2 discusses the broader role of the public in Swiss political culture in order to put its role in regulation and the various conceptions of 'the public' into perspective. It addresses the role of experts in the Swiss political system, and illustrates this discussion with examples from the Gene Technology Law's legislative process and licensing procedure. Finally, Section 6.4.3 explores the Swiss legislative culture, including both its legal traditions and its legislative approaches.

6.4.1 Democracy and Federalism

From the adoption of its constitution in 1848, the Confederation of Switzerland has drawn on two forms of government: federalism and direct democracy. Balancing these elements has contributed to the development of a consensus model of decision-making that exemplifies the elements of a consociational democracy.[48] Some scholars classify the unique Swiss model as a semi-direct democracy,[49] or as an interactionist model of direct democracy.[50] These definitions all indicate that the Swiss democracy is not a system in which the 'winner takes all'. Instead, the political system emphasizes respect for minorities. According to Fleiner, the mixture of federalism and direct democracy is a necessary condition for establishing a consociational democracy, which guarantees the freedom and

---

47 SAG-gruppe, D. Ammann and A.F. Goetschel, SAG-Studienpapiere, *Die Verfassungsnorm der Würde der Kreatur/ Konsequenzen für die Zulassung genmanipulierter Tiere,* 1998.

48 A. Lijphart, *Patterns of Democracy*, New Haven, Yale University Press, 1999, Ch. 3.

49 Linder, 'Direct Democracy', p. 110.

50 Reich, 'An Interactionist Model of Direct Democracy', pp. 1-31.

self-determination of the Cantons.[51] For the Swiss people, this political system prevents the 'tyranny of the majority' and ensures that minority interests are represented within the state. Fleiner claims that consensus democracy legitimates political and legal decisions with regard to minorities.[52] However, as this section will argue, these claims only hold for traditional minorities.

Federalism and direct democracy are institutionalized in the Swiss Constitution. The combination of these forms of government guarantees two basic principles: Equality and autonomy of the Cantons.[53] Furthermore, it ensures that all power in the Swiss Confederation comes from the people and the Cantons.[54] For the Cantons, the combination of federalism and direct democracy implies both a right and a duty. They have a right to participate in decision-making processes at the national level and at the same time a duty to cooperate.[55] The Cantons have several possibilities for initiating legislation and being involved in decision-making, such as the cantonal initiative.[56] Furthermore, the Cantons are represented in the Council of States that together with the National Council forms the Federal Assembly. Both Councils have equal rights and powers in Parliament concerning initiating draft legislation and launching parliamentary initiatives. Furthermore, both Councils have to consent to federal decisions. This system is considered one of the core elements for ensuring cantonal influence in federal decision-making.[57] Federalism together with extensive decentralization guarantees the partial autonomy of the Cantons.

Respecting the autonomy of the Cantons and their right to self-determination has far-reaching consequences. Although the Cantons differ in size, both with respect to population and territory, they have equal sovereign rights. The value of the vote of a citizen in the Canton of Zürich is, therefore, much less than that of citizen in Appenzell, for example.[58]

To ensure public involvement in decision-making, the Constitution grants the Swiss people several popular rights. In particular, the Swiss direct democracy involve a possibility to 'veto' parliamentary decisions by means of a popular referendum.[59] These referenda have other consequences as well, such as slowing down various societal and political developments.[60] In addition to bringing legislation to a popular vote, calls for popular rights are often used as a bargaining instrument by interest groups and organized political forces.[61]

51 T. Fleiner, 'Recent Developments of Swiss Federalism', in *The Journal of Federalism*, 32/2, 2002, pp. 12-13.
52 *Ibid.*, p. 17.
53 Fleiner refers to these principles as freedom and self-determination which are equivalent terms.
54 Church, *The Politics and Government of Switzerland*, p. 51.
55 A. Vatter, 'Federalism', in U. Klöti, P. Knoepfel, H. Kriesi, W. Linder, Y. Papadopoulos, P. Sciarini (eds.), *Handbook of Swiss Politics*, Zürich, Neue Zürcher Zeitung, 2007, pp. 75-77.
56 Vatter, 'Federalism', p. 83.
57 *Ibid.*
58 A. Vatter, 'Consensus and Direct Democracy: Conceptual and Empirical Linkages', in *European Journal of Political Research*, 38, 2000, p. 174.
59 Linder, 'Direct Democracy', p. 102.
60 W. Linder, *Swiss Democracy: Possible Solutions to Conflict in Multicultural Societies*, New York, St. Martin's Press, 1994, p. 99.
61 Linder, 'Direct Democracy', p. 114.

These popular rights have transformed Swiss democracy into a consensus model of decision-making characterized by negotiation, broad coalitions, pre-parliamentary procedures, and power-sharing.[62]

First of all, the composition of the Federal Council as a power-sharing executive branch ensures that the interests of traditional minorities are taken into account. The composition of the Federal Council reflects the autonomy of the Cantons. An unwritten rule prescribes that four of the seven members of the Federal Council are from a Swiss-German canton, and three members from a Swiss-French or Swiss-Italian canton. Furthermore, the Council's members are selected on the basis of their political preferences. The four major parties are proportionally represented in the Council, which explicitly reflects a system of concordance.[63] This system adopts modes of power-sharing, which can be considered a particular form of consociational democracy.

Second, a strong tradition of interest group corporatism in Switzerland has contributed to establishing a consensus model of decision-making. As Lijphart et al. argue, 'as the degree of consensus democracy increases, the degree of corporatism increases'.[64] In ideal-type consociational democracies, major social conflicts are settled through bargaining among organized groups.[65] Since the late nineteenth century, interest associations have become increasingly important for the resolution of economic and social conflicts. Consequently, a system of interest associations has emerged. The Swiss federal government has encouraged these developments. The federal government argues that incorporating organized interests into official decision-making procedures will contribute to solving social and economic conflicts. These bargaining processes are supposed to ensure broad public support.[66]

Third, institutions such as the pre-parliamentary procedure (*Vernehmlassungsverfahren)* have increased the importance of bargaining in political decision-making processes. The pre-parliamentary procedure, instituted in 1947, is a method for ensuring that the considerations of all relevant groups are taken into account at an early stage. This procedure makes consultation of various actors mandatory.[67] More importantly, the introduction of the pre-parliamentary procedure aims to avoid the threat of an optional referendum.[68] It has also influenced the political economy of referenda, which are no longer dominated by 'no-talk, just vote, drive-through' referenda.[69] Instead, the playing field is characterized by a deliberative process in which the individual voter prefers to participate in the formation of well-considered decisions.

The Swiss form of government has not ensured broad deliberation and inclusion for everyone, however. In Switzerland, as in other countries characterized by 'segmented

62 *Ibid.*, p. 113. Lijphart, *Patterns of Democracy*.

63 Lijphart, *Patterns of Democracy*, pp. 34-35.

64 A. Lijphart and M.M.L. Crepaz, 'Corporatism and Consensus Democracy in Eighteen Countries: Conceptual and Empirical Linkages', in *British Journal of Political Science*, 21/2, 1991, p. 241.

65 G. Lehnbruch, 'Consociational Democracy and Corporatism in Switzerland', in *Publius: The Journal of Federalism*, 23, Spring 1993, p. 44.

66 *Ibid.*, pp. 51-53.

67 Lijphart, *Patterns of Democracy*, p. 53.

68 Art. 147 BV.

69 Reich, 'An Interactionist Model of Direct Democracy', p. 22.

pluralism', interest associations are strongly connected to the political parties that represent the traditional minorities of the Confederation.[70] On the one hand, this leads to the formation of well-organised groups and a strong relation between political parties and their grassroots supporters. On the other hand, however, recent developments have led to questions about whether this system excludes 'new' minorities from the political process. 'New' minorities are often less well-organised, if they are organised at all. Reich coins this the 'blind spot of direct democracy'.[71] Although traditional minorities are diverse as well, together they are at least politically homogeneous.[72] The 'new' minorities do not share similar political values. The Swiss system seems to fail in protecting new minorities, since these 'new' minorities are not represented or organised in the system. Decisions presented as based on consensus do not represent the interests of 'new' minorities, as their concerns and preferences are not included in deliberation. Here again, the notion of consensus hides a lack of agreement among all actors.

These recent developments complicate the assumption of public involvement by representation. Moreover, the political system again risks falling prey to the tyranny of the majority, which was supposed to be avoided by the mixture of elements of both direct democracy and federalism.[73]

6.4.2 The Role of the Public

The Swiss people play an active role in political and legal decision-making. Section 6.3.1 argued that the Swiss acknowledge two conceptions of 'the public'.[74] This section begins by exploring the various popular rights that provide an official role to the public as one actor. The second conception of the public, which conceives of the public as represented by a variety of interest groups and organizations, was extensively discussed in the previous sections. To avoid repetition, this section restricts itself to discussing popular rights and the role of experts.

Popular Rights

The three pillars of Swiss direct democracy are built on the Swiss popular rights: the popular initiative,[75] the obligatory referendum,[76] and the optional referendum.[77]

---

70 Vatter, 'Federalism', p. 86.

71 Reich, 'An Interactionist Model of Direct Democracy', pp. 26-29.

72 Fleiner, 'Recent Developments of Swiss Federalism', pp. 5-7.

73 Lehnbruch, 'Consociational Democracy and Corporatism in Switzerland', p. 56.

74 See Church, *The Politics and Government of Switzerland*, p. 56; see also Chapter 4 of this book.

75 Art. 139 BV; an initiative of the people requires the signature of 100.000 citizens and it is held for a partial amendment or for a total revision of the constitution. The signatures need to be collected within a period of 18 months and after that both the Federal Council and the Federal Assembly debate the proposal and present their recommendations to the voters. Linder, *Handbook of Swiss Politics*, pp. 105-106.

76 Art. 140 BV, an obligatory referendum is approved if it reaches a majority of the voters and at least 12 votes of the 26 cantons.

77 Art. 141 BV, an optional referendum is launched if 50.000 citizens or eight cantons ask for a referendum within three months.

In a popular initiative, 'the people' can initiate a request or a specific proposal for partial amendment or revision of the constitution. Despite their great popularity, these iniatives are seldom accepted: between 1892 and May 2008 162 initiatives were launched and only 15 of these were approved.[78] Instead, in most cases, the parliament responds by launching a counter-proposal, as they did in the case of the Gen-Lex motion, which was a counterproposal against the initiative for a general ban of GMOs. These counter-proposals by the parliament are often successful.

Nevertheless, it would be a misreading of the popular iniative to claim that it has no effect on federal decision-making. These initiatives trigger political processes, as reflected in the various parliamentary counterproposals.[79] These proposals often pave the way for new political trends and have brought about some important changes in the Swiss political system, such as proportional representation and a provision on equality.[80] At the same time, they have also led to controversial regulations. In 2009, for example, a popular iniative resulted in a ban on the building of minarets.[81]

The second popular right, the obligatory referendum, ensures that the Swiss people have an official say in decision-making. All amendments to the Constitution as well as ratifications of treaties involving membership in organisations of collective security or supranational bodies require a double majority: both the people and the cantons have to approve. It is often said that the obligatory referendum slows down social and political developments: initial attempts at regulation often fail, as voting against an amendment is used to express concerns. Consequently, important changes need time.[82] Nevertheless, obligatory referenda ensure that the people have a say in important decisions. Moreover, this instrument, which is regularly used every three to four months, ensures broad public debates concerning diverse political issues.[83]

The third popular right is the optional referendum. Parliamentary decisions such as federal statutes that are generally binding, and permanent international treaties on membership in international organisations or multilateral legal harmonisation can be, but do not necessarily need to be, put to a popular vote. In reality, only around 7% of bills are submitted to a popular vote. Therefore, the influence of the optional referendum in decision-making is said to be limited. It is often argued that this instrument is not very effective

---

78 Reich, 'An Interactionist Model of Direct Democracy', p. 19.

79 Iniatives are often considered a tool for progressive political organizations.

80 Examples can be found in Linder, 'Direct Democracy', p. 112.

81 Federal Parliament, resolution on the admissibility of the initiative (*Bundesversammlung. Bundesbeschluss/ Assemblée Fédérale, arrêt federal*) Bundesblatt 2009.

82 Linder, 'Direct Democracy'. During the period between 1848 and 2005, 205 referenda took place and around 75% of them were approved. Nonetheless, Linder argues that since still 25% is rejected, it is obvious that due to the obligatory referendum the development of the modern welfare state, the centralisation of the tasks of the government, and several constitutional decisions related to the extension of electorate came about so late. Slowing down developments is also influenced by the relatively large influence of small cantons. Small cantons are most often more conservative than the bigger ones. However, Cantons have equal sovereign rights. Consequently, small cantons 'can organize a veto to block democratic majorities' (Linder, *Swiss Democracy*, p. 159) and thus progressive developments.

83 Reich, 'An Interactionist Model of Direct Democracy', p. 14.

for the individual voter. Instead, it is public and private organized groups, cantons, and federal institutions that are able to effectively make use of optional referenda.[84]

In the optional referendum, we can recognize both conceptions of the public. First of all, the optional referendum uses a popular opposition as a control mechanism. If 'the people' are not in favour of a proposal, this instrument ensures that they can put it to a vote. Although this right is addressed to the general public, citizens are more successful if represented by interest organisations. These organisations are more effective than individual voters, as they have more bargaining power in political decision-making, and more financial and organisational ability to call for a referendum. Calling for an optional referendum requires 50,000 votes to be gathered within 100 days.

Together, these popular rights ensure that the Swiss people have an official say in decision-making. However, these rights cannot stand on their own. They also require that the Swiss people be informed about the issue on which they have to vote. The Swiss system can only function if the Swiss government ensures the broad provision of information. In Switzerland, this involves the organization of various public debates in the media and at schools; the distribution of mailings about the issues at stake; and the discussion of the issues in newspapers and on local and national news programs. This makes the Swiss direct democracy an interactive model. Information enhances deliberation about the political issues that are put to vote.[85] The individual voter can make a well-considered decision by participating in debates. Additionally, the provision of information also ensures political transparency. Transparency makes Switzerland a direct democracy as well as a federal state. Transparency ensures that traditional minorities feel acknowledged in decision-making, and ensures the legitimacy of the decision-making system.[86]

The Role of Experts in Decision-Making

Experts play an important role in Swiss decision-making processes. First of all, organized groups participate in decision-making as 'expert' representatives of the people. Second, experts can have an advisory role, such as that played by the SECB and the ECNH in the GTL's licensing procedure. Third, experts can be involved in stimulating and organising public debate. In reality, the last two functions are often combined. The SECB and the ECNH, for example, are directed to advise the official institutions in their field of expertise as well as to inform the public.

The ECNH's role in informing the public and being engaged in the public debate is controversial for various reasons relating to the ECNH's interpretation of its mandate. To begin with, the ECNH's members considered 'being engaged in public dialogue about

84 Lehnbruch, 'Consociational Democracy and Corporatism in Switzerland', p. 54.

85 Reich, 'An Interactionist Model of Direct Democracy', p. 24.

86 L.P. Feld and G. Kirchgassner, 'Direct Democracy, Political Culture, and the Outcome of Economic Policy: A Report on the Swiss Experience', in *European Journal of Political Economy,* 16, 2000, p. 292. Feld and Kirchgassner do not discuss the transparency of decision-making, but instead underline the minority position which is protected in the Swiss system and the role of information for the feeling of belonging.

moral issues' as primarily a task of informing the public. The members did not feel obliged to participate actively in public debate. During interviews, it became clear the ECNH's members considered their public tasks as subsidiary to their advisory role.[87] Moreover, the members stated that lay peoples' viewpoints were irrelevant for their official statements, as well as for their advice to official institutions on licensing. Instead, they indicated that only rational deliberation was important for adequately performing their task.[88] Participating in the public debate on a horizontal level was therefore deemed unnecessary; it was sufficient to provide information to the public and stimulate public debate by feeding it.[89]

This interpretation in itself would not be problematic were it not for the fact that the federal offices saw the ECNH's statements as a substitute for public debate.[90] Consequently, the semblance of consensus presented by the committee excluded certain voices that would normally be expressed in public debate.[91] This was corroborated by the fact that the members considered public opinion irrelevant to their statements. The ECNH does not aim for inclusion. At the same time, there is no other official channel for including public voices in decision-making. Public debate was stimulated only by private organisations outside the political and legal arena. We could question whether the ECNH should play a more active role in public debate, or whether public input should be incorporated via other tracks.[92]

These conclusions about expert involvement together with the exploration of popular rights put the analysis of public involvement into perspective. Swiss popular rights involve only the acceptance or approval of the people, and do not imply any public input. Instead, public input is established through the representation of individuals by interest groups and expert committees as well as through public debate. The effectiveness of these methods, however, is doubtful.

6.4.3 *The Legislative Culture*

Despite the fact that the first article of the Swiss Civil Code requires that judicial practice recognize doctrinal writings, Swiss legislative culture is not dominated by legal doctrine.[93] Explicit references to legal theories are not prominent in the Swiss legal tradition and legislative culture. Nevertheless, in Swiss legal decision-making processes we can recognize a positivist view of law.

Societies' legal traditions are related to the legal families to which their legal systems belong, and the legal theories that have influenced their conceptions of law. Swiss legal

87 Personal communication with ECNH members, Spring 2007.

88 K.P. Rippe, 'Ethikkommissionen als Expertengremien? Das Beispiel der Edgenössischen Ethikkommission', in K.P. Rippe (ed.), *Angewandte Ethik in der pluralistischen Gesellschaft*, Fribourg, Freiburger Universitätsverlag, 1999, pp. 359-370.

89 Grotefeld, 'Wie wird Moral ins Recht gesetzt?', in *Archiv. für Rechts- und Sozialphilosophie*, 2003, p. 315.

90 Bovenkerk and Poort, 'The Role of Ethics Committees in Public Debate', p. 30.

91 *Ibid.*

92 Bovenkerk and Poort, 'The Role of Ethics Committees in Public Debate', pp. 30-31.

93 Art. 1 ZGB.

history is rather diverse, since the Cantons have legal roots in German and French legal systems. Both of these legal systems belong to the civil law tradition, which is consequently most influential.[94] The legal traditions of the Cantons influenced the 1848 Constitution of the Modern Swiss Confederation.[95] Most prominently, the Swiss legal system was influenced by the German civil law system. For example, the Swiss Civil Code is oriented to the German Civil Code.

Very few Swiss works have appeared in the field of legal philosophy. However, the influence of German and Austrian legal positivist philosophers is prominent.[96] According to Reich, the supremacy of the Constitution shows the implementation of Kelsen's philosophy of the hierarchy of law.[97] The Constitution is the basis of the Swiss federal and legal system. The strong emphasis on the Constitution exemplifies the influence of legal positivism in Swiss ideas of legal validity.

Grotefeld argues that the validity of moral concepts in a legal framework should also depend on public acceptance, especially in a semi-direct democracy such as Switzerland.[98] Thus, validity does not necessarily imply a moral substantive justification. Here Grotefeld adopts Rawls' idea that: 'Our exercise of political power is fully proper only when it is exercised in accordance with a constitution the essentials of which all citizens as free and equal may reasonably be expected to endorse in the light of principles and ideals acceptable to their common human reason.'[99] Moral legitimacy should be found in the approval of the public, he argues, and therefore cannot be fully neutral towards concepts of the good life. Thus, the legislature cannot be fully neutral by weighing the concerns and preferences of the public.

Here, we can recognize again an emphasis on popular approval, which brings us back to the prominent role of popular legitimacy in the Swiss legal system. Popular legitimacy justifies the incorporation of morality into the legal framework in a legal system influenced and characterized by a positivist view of law.

The Swiss legislature generally adopts a product-based legislative approach.[100] In a product-based approach, the regulatory framework presents a 'product' consisting of clear and concrete norms. The product-based approach corresponds with two traditional functions of law: the instrumental and the constitutional functions. Legislation serves an instrumental function by ensuring fair processes in which both the public and the Cantons have a say. And it serves a constitutional function by acting as an instrument for constituting fundamental values.

The legislature acknowledges that some kind of issues require a different approach. For complex issues such as animal biotechnology, they have determined that a process-based approach is more adequate. This approach regulates the process for dealing with future issues rather than aiming to formulate concrete norms. At first sight, this process-based

94 L. Carlen, *Rechtsgeschichte der Schweiz*, Bern, Francke Verlag, 1988, Ch. 3 and Ch. 9.

95 *Ibid.*, pp. 93-94.

96 C. Szladits, *A Guide to Foreign Legal Materials*, Dobbs Ferry, Oceana Publications Inc., 1959, pp. 492-493.

97 Reich, 'An Interactionist Model of Direct Democracy', p. 31.

98 Grotefeld, 'Wie wird Moral ins Rechts gesetzt?', pp. 304-312.

99 J. Rawls, *Political Liberalism*, New York, Columbia University Press, 1996, p. 137.

100 Errass, *Öffentliches Recht der Gentechnologie im Ausserhumanbereich*, pp. 112-115.

approach seems to have much in common with the ideal-typical model of the interactive legislative approach. This approach acknowledges the need for open norms and an ongoing process of concretisation.[101] Nevertheless, despite these similarities between the process-based legislative approach and the interactive legislative approach, several fundamental differences can be observed.

First of all, Switzerland's reasons for using a process-based approach differ from the ideal-typical model of the interactive legislative approach. The Swiss use of open norms can be explained by a lack of consensus. about concrete norms on the one hand and the need for regulation in the short run on the other. However, rather than seeing the process-based approach as a superior alternative for regulating issues with complex moral implications, the Swiss legislature views the process-based approach as a second-best option. The process-based approach is therefore unstable, and risks falling back on a more traditional product-based approach as soon as possible. As a result, a dynamic process of further norm development has not been established in Swiss legislative practice. Here, the influence of positivism can be observed: open norms obstruct legal security, and therefore the law should be clear and concrete. Second, legal positivism underlines the regulatory character of law, not its symbolic or communicative functions. Here, the use of open norms is not derived from their potential to enhance the symbolic or communicative functions of law, but is rather the result of more pragmatic reasoning. The positivist Swiss conception of law and legislative functions clarifies its difficulties concerning the implementation of open norms. These norms are not seen as symbolic ones inviting further communication, but instead as guidelines for concrete action.

6.5 Conclusion

Switzerland seemed at first sight to present fertile ground for an interactive legislative approach. The Swiss mixture of federalism and direct democracy, the semi-direct democracy, the interactive model of direct democracy, the consensus model of decision-making, expert involvement, and the incorporation of fundamental values into the legislative framework of the GTL all contribute to this impression. These elements of the Swiss political system create outstanding possibilities for interactive and dynamic processes of legal norm development. Decision-making processes are characterized by broad participation in which actors deliberate on a horizontal level. Furthermore, the dynamics of the interactive process seem to be reinforced by the incorporation of fundamental values in legislation and the involvement of expert committees.

Despite the existence of this structure, however, Swiss practice did not correspond with the model interactive legislative approach with respect to legislation on animal biotechnology. To start with, the process of norm development stagnated. This is exemplified by the failure to further develop the concept of 'dignity of living beings'. The concept of 'dignity of living beings' was developed in application, but not further analyzed in its essence. The semblance of consensus presented by the ECNH stifled further development, since norm

101 *Ibid.*; Grotefeld, 'Wie Wird Moral ins Rechts gesetzt?', p. 302.

development is less urgent if a consensus already exists. Existing norms were considered concrete guidelines for action instead of symbolic norms in need for further development.

The level of interaction in the legislative process also proved inadequate. To begin with, the horizontal level of decision-making needs to be put into perspective. The role of the public is not explicitly horizontal, but merely indirect, operating via representation by organised interest groups. At first sight this does not seems to be problematic since the Swiss people are very well-informed and involved. Popular rights allow the Swiss people to control the legislature. At the same time, recent developments have highlighted the blind spot of democracy. The organised interest groups and well-informed involved citizens mainly represent the traditional minorities. 'New' minorities are not included in the political system, since these minorities are not organised into interest groups or represented in traditional political parties.

Interaction is also doubtful since the main incentive for public involvement is increasing public support. As a result, this process often chooses the middle course of satisfying the greatest number of actors possible. This is then presented as if it were a consensus, which in reality often lacks substantiation or is indeed not consensus at all. This also results in the exclusion of certain groups.

The inadequacy of norm development and the level of interaction in the regulation of animal biotechnology may be partially explained by the Swiss emphasis on popular legitimacy in terms of public support, rather than in terms of public input and inclusiveness. Further explanation is found in Swiss legal traditions, which privilege product-based over process-based legislative approaches. Although Switzerland adopted a process-based approach for regulating animal biotechnology, the preference of the Swiss legislature for product-based approaches turned the process-based approach into a high risk one that fell back on traditional structures and strategies. Ultimately, symbolic and communicative legislative functions are less dominant than the regulatory function of law.

Despite its initial promise, therefore, the Swiss political system does not seem to provide sufficient conditions for an interactive legislative approach to function adequately. This outcome provides interesting insights into the potential pitfalls faced by interactive legislative processes.

First of all, it can be questioned whether consensus should be a leading aim in establishing a dynamic process. Swiss political culture depends on consensus-based decision-making processes. As we have seen, a strong drive for consensus leads to some disadvantages. Consensus-driven models can easily exclude social groups who are not represented in the bargaining process, such as the 'new' minorities. In addition, the strong drive for consensus runs the risk of obscuring essential questions as well as controversial viewpoints, as occurred in the elaboration of the dignity of living beings. Finally, consensus-based decisions are difficult to justify substantively. Often, consensus either lacks substantive grounding, or presents a partial opinion as based on consensus in order to satisfy the majority.

Second, it can be questioned whether the blind spot of democracy represents the limit concept of an interactive legislative approach, as it seems to restrict the extent of involvement by supposing a politically homogenous people. These questions will be further analyzed in Part III of this research.

7 The Regulation of Animal Biotechnology in the Netherlands

7.1 Introduction

This chapter presents the third and final case study for analysing the interactive legislative model. This case study explores the Dutch regulation of animal biotechnology and its legislative process. The Dutch regulation of animal biotechnology and the Dutch political culture seem at first sight to contain characteristics of the ideal-typical model of the interactive legislative approach. Dutch political culture is characterized by a broad pluralism. This builds on a long tradition of segmented pluralism, which is well known as the '*verzuiling*', or 'pillarization'. The Dutch pillars are traditionally divided according to political as well as religious differences. In the political system, this '*verzuiling*' resulted in a politics of accommodation. In a more recent incarnation, this pluralism has taken the form of the '*poldermodel*',[1] which entails a culture of negotiation and bargaining. Dutch government functions by means of a coalition with a majority cabinet. Although both the '*verzuiling*' and the '*poldermodel*' have slowly begun to vanish, their influence on political culture is still prominent. Participatory methods, involvement of experts, and a culture of bargaining are the heritage of these traditions. These characteristics can also be recognized in the regulation of animal biotechnology, which prescribes a licensing procedure that is open to participation by various actors.

This chapter follows the same structure as Chapter 5 and 6. Section 7.2 begins by examining the Dutch regulation of animal biotechnology. It explores both the legislative process and the regulation in practice. Second, Section 7.3 discusses public involvement in legislation, the role of moral values in legislation, and the legitimacy of legal norms. Section 7.4 then turns to the democratic and legal traditions of the Netherlands. This discussion focuses on assessing the Dutch legislative approach and its conformity with the interactive model of legislation. This section also examines the Dutch culture of broad participation. Finally, Section 7.5 formulates a preliminary answer to the question of whether the interactive legislative approach was able to function adequately in the Netherlands. These remarks will be discussed more thoroughly in Chapter 8 and in Part III of this research.

---

1 Stichting van de Arbeid, 'Centrale aanbevelingen inzake aspecten van een werkgelegenheidsbeleid', (Akkoord Van Wassenaar), 1982.

THE INTERACTIVE LEGISLATIVE APPROACH IN THE NETHERLANDS

7.2 THE REGULATION OF ANIMAL BIOTECHNOLOGY

In the Netherlands, animal biotechnology is regulated by the Animal Health and Welfare Law (the Law), which came into force in September 1992. The Law creates a regulatory framework that governs various topics concerning the health and welfare of animals. In general, it adopts a 'no, unless' policy.[2] Chapter 4 of the Law sets the basis for regulating animal biotechnology. This basic law requires further elaboration by administrative orders. Rules regarding the licensing of biotechnological procedures involving animals are set forth in the Animal Biotechnology Decree (the Decree), which came into force in 1997.[3]

This section discusses the Law and the Decree in two parts. First, Section 7.2.1 examines the long and complex legislative procedure that led up to the adoption of these regulations. Second, Section 7.2.2 explores the regulations in practice by examining the Dutch licensing procedure for animal biotechnology.[4]

7.2.1 The Legislative Procedure

A first tentative step towards protecting the health and welfare of animals was taken in 1981, with the publication of the Memorandum on Government and Animal Protection *(Nota van Rijksoverheid en Dierenberscherming)*. This memorandum explicitly recognized the intrinsic value of animals, and acknowledged the government's role in protecting them. This memorandum immediately resulted in a first draft of the Animal Health Law.[5]

Over the next several years, the draft Animal Health Law went through various changes. 1985 was an especially eventful year. First of all, in 1985 the concept of animal welfare was introduced onto the political agenda. As a result, the name of the draft regulation was changed to the Animal Health and Welfare Law. Second, in 1985 the Animal Protection Organisation proposed a stricter alternative law. Although the proposal was not accepted by Parliament, it influenced political thinking. A number of political parties favoured this stronger alternative and argued for a stricter policy regarding the protection of animals. Others feared that the Animal Protection Organisation's proposal would lead to over-regulation.[6] Neither the draft Animal Health and Welfare Law nor the proposed

2 Explanatory Memorandum of the Animal Health Welfare Law, 1992.

3 Explanatory Memorandum of the Animal Biotechnology Decree, 1997.

4 One remark; recent developments are not included in this research, as the CAB does not operate any longer in legislative practice. This study involves function of the Decree until July 2009. After July 2009 various changes in legislative practice are made. I will not ignore this changes, but these will play a less central role. Nevertheless, these changes shine a interesting light on the conclusions of this research and will therefore at least be referred to where relevant.

5 First draft, Animal Health Law, Parliamentary Documents, Second Chamber, 1980-1981, 16447, nr. 3.

6 Parliamentary Documents, Second Chamber, 1988-1989, 16447, nr. 15.

alternative, however, addressed animal biotechnology. Animal biotechnology and specifically its moral aspects did not come onto the political agenda until the late 1980s, when society began to show increasing interest in the consequences of this new technology and demanded regulations in the context of human obligations towards animals.[7] In 1986, a report about the consequences of animal biotechnology was published on behalf of the CPJ, an agrarian youth organisation (*Culturele Plattelands Jongeren*). This report recommended a more thorough discussion of the moral aspects of animal biotechnology, and argued that biotechnological procedures with animals should be prohibited, unless they could contribute to improving health care.[8] The Minister of Agriculture, Nature and Food Quality (Minister of Agriculture) responded by asking the NOTA (*Nederlandse Organisatie voor Technologisch Aspectenonderzoek)*[9] to analyze the moral considerations of the genetic modification of animals from a technological perspective.[10] The Minister of Agriculture also appointed an Advisory Committee on Ethics and Animal Biotechnology (*Commissie van Advies Ethiek en Biotechnologie bij Dieren),* known as Schroten I (after its chairman), in 1989. Schroten I presented a report in May 1990. It recommended a licensing procedure based on a 'no, unless' policy, which would employ an ethical review of all biotechnological procedures involving animals.

In the meantime, two companies, Gene-Pharming Europe BV and IVO, were experimenting with cows that could be induced to produce human proteins in their milk. This resulted in the birth of Herman, a transgenic bull, in 1990. The bull's female offspring would be able to produce milk with human lactoferrin. This milk could serve as a basis for creating medicines to treat intestinal infections.[11] Herman's birth met with a great deal of resistance in political and public circles, and among animal protection organisations. Further procreation by Herman was (temporarily) prohibited.[12] Under pressure from various interest groups and public oppression, a parliamentary debate about the moral acceptability of animal biotechnology ensued. The Animal Protection Organisation initiated research into the moral aspects of animal biotechnology, in particular the research of Gene-Pharming and IVO.

All of this media, public and political attention eventually resulted in the development of a unique set of rules, which required the explicit review of biotechnological procedures with animals. In 1990, the Minister of Agriculture presented a new proposal that supplemented the draft Animal Health and Welfare Law with a chapter on animal biotechnology. In its memorandum of amendment, the government acknowledged the recommendations of Schroten I and adopted the 'no, unless' policy.

7 Parliamentary Documents Second Chamber, 1988-1989, 16447, nr. 16.

8 M. Linskens, *Transgene landbouwhuisdieren: Het overwegen waard?*, Wageningen, Nederlandse Christelijke Plattelandsjongeren/ Wetenschapwinkel LUW, rapport 31, 1989, p. 1.

9 now Rathenau Instituut.

10 L. Fretz and L. Sterrenburg (red. A. Rijnen), *Het maakbare dier geregeld: Politieke besluitvorming en genetische modificatie van dieren: Een rapportage aan het Parlement*, Den Haag, NOTA, 1993.

11 H. Schellekens, *De DNA-Makers*, 1993.

12 F.W.A. Brom, *Onherstelbaar verbeterd.*

This new policy indicated that biotechnological procedures with animals were in principle prohibited. The Minister of Agriculture emphasized that there was no consensus on the moral impact of animal biotechnology, and there remained uncertainty about its consequences. As a result, the Minister determined that basic rules should be set and further crystallized in the future. At the same time, broad public and political debate on the draft law did not ensure an easy process. On the contrary – it took another year before the draft was accepted, and another four before the Decree came into force. In that time, members of Parliament put forward several amendments proposing more specific rules for regulating animal biotechnology akin to the recommendations of Schroten I.[13] The Van Noord and Swildens-Rozendaal amendment specifically focused on bringing into effect the 'no, unless' policy. They complemented the draft with a licensing-procedure in which the moral considerations and the consequences for health and welfare of the animals would be balanced against the value of the proposed research.[14] This amendment was finally accepted by Parliament on 17th of April 1991.

Some in Parliament questioned the extent of the 'no, unless' policy. Furthermore, they debated the Law's conceptualisation of intrinsic value and the ethical review system of licensing. The Parliamentary members agreed, on the one hand, that intrinsic value may outweigh economic, educative, and recreational interests. But on the other hand, they also recognized that applications for medical research could justify the use of these technologies. As a result, they agreed to appoint an ethics committee – the Committee on Animal biotechnology (CAB) – to clarify and strengthen norms on the moral status and position of animals.[15]

The CAB is a committee of experts from various fields, which is explicitly directed to reflect on the moral aspects of biotechnology. The president of Schroten I was reappointed as president of the CAB. The CAB could therefore easily adopt Schroten I's earlier recommendations on the moral impact of biotechnological procedures involving animals.

Finally in September 1992, the Parliament accepted the draft and implemented the Law. Under Article 66 subsection 1, animal biotechnology permit applications were to be lodged with the Minister of Agriculture, and would be granted when a) the procedures had no unacceptable consequences for the health and welfare of the animals, and b) there were no moral objections to the procedures. In the meantime, the provisional CAB[16] discussed its first case: the procreation of the transgenic bull Herman, whose first female offspring had been born earlier that year.[17]

---

13 For example, Beckers-De Bruijn which was rejected. Parliamentary Documents, Second Chamber, 1990, 16447, nr. 79.

14 Parliamentary Documents, Second Chamber, 1988-1989, 16447, nrs. 79-81.

15 Explanatory Amendment of the Animal Health and Welfare Law.

16 At this time the CAB was not yet fully installed. The provisional body that heard this case was known as Schroten II (after its chairman).

17 Voorlopige commissie ethische toetsing genetische modificatie van dieren (Schroten II), *Advies inzake het dossier 'Weefselspecifieke expressie van genen in de melkklier van genetisch gemodificeerde runderen'*, Den Haag/Utrecht, LNV, 1992.

Animal biotechnology is further regulated in the Animal Biotechnology Decree. Since the main issues had already been dealt with during the development of the Law, the Decree's legislative procedure mainly involved the explication and practical application of these issues. The main topics of discussion related to the framework for ethical review, the CAB and its terms of reference, the field of application of the Decree, and the import of transgenic animals.[18] During these discussions, both the public and scientists could and did express their concerns.

In 1993, NOTA initiated a public meeting in which a panel of fifteen informed citizens posed questions about the genetic modification of animals.[19] This meeting was organized in three phases. In the first phase, experts and interested parties clarified the various aspects of genetic modification. During the second phase, the panel's members and the general public had the opportunity to ask questions to the experts and interested parties. The last phase ended with a statement on animal biotechnology and a closing discussion.[20] Most members of the panel agreed on a temporary moratorium for genetic modification. The members also called for a further examination of alternatives, noting that alternatives that prevent research from harming nature and avoid using animals should be preferred. Finally, the panel called for further public debate.[21] This was felt to be especially important since the government had not yet decided on a moral position.[22] The outcomes of this public meeting were not incorporated in the animal biotechnology regulations. However, they indicated the direction of public sentiment at the time.

Scientific experts on animal biotechnology also discussed the draft Decree. These experts feared the licensing procedure would slow down further scientific developments, especially since licensing was also required for biotechnological procedures involving mice and rats, animals that are often used in biomedical research. Procedures involving mice and rats were already reviewed by other committees, such as the Animal Experiments Committees and the Committee for Genetic Modification. An additional review would merely slow down the procedure. The concerns of these experts increased after the CAB suggested a 'concept application' form of licensing. They also raised various questions concerning the concept of the intrinsic value of animals. Responding to these concerns, the committee on medical science of the Royal Dutch Academy of Science presented a report entitled 'Public Health and Biotechnology' in 1995. This committee suggested that almost 95% of transgenic research should be excluded from the requirements of the licensing procedure. This report received a lot of attention in political circles and among scientific experts. As a response, in May 1996 the Minister of Agriculture organized a meeting between various stakeholders in the field of biotechnology to give their concerns a voice.

18 Explanatory Amendment of the Animal Biotechnology Decree.

19 The public meeting is a lot like the consensus conferences as organized in Denmark. However, this meeting was not focused on establishing consensus as such. It only had a similar structure and similar incentives concerning public input.

20 Fretz and Sterrenburg, *Het maakbare dier geregeld.*

21 *Ibid.*, p. 36.

22 *Ibid.*, p. 35.

To a certain extent, this meeting resulted in an agreement about the licensing procedure.[23] Stakeholders proposed the merging of biotechnology procedures with comparable moral impacts into a single so-called 'unit' for review.[24]

Further elaboration of the review framework was left to a committee of experts known as the Brom Committee, after its chairman. The Brom Committee presented its report to the CAB in September 1996. This report listed a number of criteria that should be used in the review framework for licensing, and detailed some recommended steps that an optimal review would contain. It outlined the system to be used, while leaving further elaboration to the CAB in the licensing procedure.[25] The Animal Biotechnology Decree was accepted by Parliament on 9th of December 1996, and came into force on the 1st of April 1997. This law sets rules concerning the composition of the CAB, its terms of reference, and the regulation applied to the licensing procedure.

It is noteworthy that neither the legislative process nor the parliamentary and societal debates were concerned with the Decree's substance, but rather focused on the conditions under which the Decree should function in the licensing procedure. This exemplifies the strong influence of experts in regulating this complex issue. Furthermore, we can recognize the Minister of Agriculture's preference for a clearer set of standards and his inclination toward an instrumental approach to law.

7.2.2 The Legislation in Practice

The 'no, unless' policy entails a restrictive use of biotechnological research involving animals.[26] In principle, research involving animal biotechnology is prohibited. However, licenses may be granted for some important societal interests. This was meant to be a provisional policy, in force until opinions on the use of these technologies had further crystallized. A more definitive stance could only be justified in the long term if no new arguments came up for consideration. The objectives of the government were: 1) the clarification of the position of animals, and 2) opinion-forming on developments in the field of animal biotechnology and its moral aspects. Here we can recognize an incentive for 'dynamics' in terms of ongoing norm development. Brom has argued that the Dutch policy regarding animal biotechnology was in fact also a policy intended to develop public morality.[27] The CAB plays a central role in this policy and in the government's strategy for establishing its objectives.[28]

---

23 Parliamentary Documents, Second Chamber, 1995/1996, 24400 XIV, nr. 57.

24 Brom, *Onherstelbaar verbeterd.*

25 F.W.A. Brom, M.T. Hilhorst, R.H.J. ter Meulen, and J.M.G. Vorstenbosch, *Het toetsen van biotechnologische handelingen bij dieren. Rapport van een commissie van externe deskundigen ten behoeve van de Commissie Biotechnologie bij Dieren,* 1996.

26 Explanatory Amendment of the Animal Health and Welfare Law.

27 Brom, *Onherstelbaar verbeterd,* Ch. 11.

28 Paula, *Ethics Committees*, Ch. 1.

During the licensing procedure, a scientist submits an application to the Minister of Agriculture, who announces this application in the *Staatscourant* and sends it to the CAB. The CAB formulates its recommendation based on an ethical review. After receiving this recommendation, the Minister draws up a draft decision. This draft decision is opened for public objections. The public can then explain their objections in a public consultation procedure organized by the Minister. In reality, it is the CAB that chairs these hearings. After considering the public comments, the CAB advises the Minister again. Officially, the Minister then decides whether or not to grant the license. In the licensing procedure, however, the Minister almost always adopts the advice of the CAB.[29] Thus, in reality, it is the CAB that has the decisive role. The final decision of the Minister is open to appeal, which can be filed at the Board of Appeal for Trade and Industry.

The Brom Committee began to develop its review system in 1996, in preparation for the instalment of the CAB. In its report, the Brom Committee listed criteria for review and recommended a step-by-step approach to licensing. This resulted in the development of a review system based on the following questions:[30]

1. Is the objective of the research of substantial interest?
2. Do realistic alternatives to the use of animal biotechnology exist?
3. Does the research pose unacceptable risks of harm to health and welfare?
4. Does the research involve an unacceptable violation of animal integrity?

Since these criteria were not yet completely crystallized, the CAB had a hand in developing the substantive arguments that would influence decisions within this framework. In its monthly meetings, the CAB discussed these criteria in light of the applications and aimed to establish consensus on the content of the review system.[31] The CAB's work and the application of the Decree to the licensing procedure have been evaluated several times.[32] These evaluations were prescribed in the Decree, and underline the temporary character of the Dutch policy. They also provide insight into the establishment of the CAB. The CAB was assigned three tasks: 1) clarification of the moral position of animals in view of biotechnology; 2) strengthening animals' moral position in view of biotechnology; and 3) identification, formulation, and assessment of problematic developments at an early stage to assist and stimulate public discussion about animal biotechnology.[33] The CAB was only partially successful in performing its tasks.

On the one hand, it has been argued that the review framework proved its usefulness in the licensing procedure, since in most cases the CAB was able to come to a unanimous decision about the moral aspects of licensing applications.[34] In the evaluation of the CAB

29 Except for one case, see Bovenkerk, *The Biotechnology Debate*, p. 179.

30 Brom et al., *Het toetsen van biotechnologische handelingen bij dieren.*

31 Bovenkerk, *The Biotechnology Debate*, p. 180.

32 See the evaluations of 2000, 2001, and 2005.

33 L.E. Paula, 'Biotechnologie bij dieren ethisch getoetst? Een onderzoek naar het functioneren van het Besluit Biotechnologie bij Dieren', Den Haag, Rathenau Insituut, 2001.

34 A. Meijer, G.K. Pikker, M.J.W.A. Schiffelers, A.M.J. van der Spek, *Evaluatie van het Besluit Biotechnologie bij Dieren*, 2005, p. 13.

it became clear that about 90 per cent of its recommendations were based on a unanimous decision. According to Meijer et al., these results reflect the effectiveness of the review framework.[35] Furthermore, the CAB established 'substantial' consensus on applications regarding a variety of research.[36] In many issues, the CAB reached consensus on both the problem definition and a solution for the problem. It seems that processes of fact-finding and constructing norms contributed to the development of a set of well-functioning practical standards[37] and concrete norms.[38] To that extent, the CAB has fulfilled its role in stimulating norm development.

However, on the other hand, the CAB's deliberations were restricted to individual cases. As a result, norm development was not completed at a higher aggregation level. It has not resulted in a set of general standards concerning the moral acceptability of biotechnology. Furthermore, the CAB's approach gives rise to some fundamental questions about the basis for regulating animal biotechnology. First of all, almost 90% of all applications were based on a unanimous 'yes' vote, which does not seem to correspond with the ideals of the 'no, unless' policy. Instead, the CAB seems to have adopted a 'yes, if' policy.[39] This undermines the symbolic value of the law. Second, the CAB has contributed little to the development of social morality. Bovenkerk correctly questions whether this 90 per cent rate of unanimous decisions reflects true consensus or whether this 'means that the more divisive moral problems were excluded from Committee discussions?'[40]

Discussions on the moral acceptability of animal biotechnology seem to have reached their saturation point.[41] However, the members of the CAB have suggested several mechanisms to break this stalemate.[42] The CAB initiated a round of four discussion meetings organised in 2002, 2003, and 2004 about general questions concerning animal biotechnology and the moral status of animals.[43] Although they enlisted a broader audience, the meetings were not successful in deepening the debate.[44]

The evaluations of the CAB and the Decree point to various explanations for the stagnation of the moral debate. First of all, it is often argued that the discussion of moral

35 Meijer et al., *Evaluatie van het Besluit Biotechnologie bij Dieren.*

36 Interview with E. Schroten (Chairman of the CAB), Spring 2009, Meijer et al., *Evaluatie van het Besluit Biotechnologie bij Dieren.*

37 Paula, *Ethics Committees*, Figure I.11, p. 77.

38 Interview with E. Schroten 2009.

39 Paula, 'Biotechnologie bij dieren ethisch getoetst?'.

40 Bovenkerk, *The Biotechnology Debate*, p. 187.

41 Paula, 'Biotechnologie bij dieren ethisch getoetst?', Ch. 3.

42 Bovenkerk and Poort, 'The Role of Ethics Committees', pp. 1-35.

43 T. Visak and F. Meijboom, 'Integriteit van dieren: Bouwsteen of Struikelblok? Verslag van de 1ste discussiebijeenkomst over biotechnologie bij dieren', Utrecht, Centre for Bioethics and Health Law, 2002; T. Visak and F. Meijboom, 'Op zoek naar alternatieven – maar voor welke doelstelling?', Utrecht, Ethics Institute, 2003a; T. Visak and F. Meijboom, 'Het maatschappelijk belang van een doelstelling: Zoeken naar goede argumenten. Verslag van de derde discussiebijeenkomst over biotechnologie bij dieren', Utrecht, Ethics Institute, 2003b; T. Visak and F. Meijboom, 'De waarde van een machtsvrije dialoog voor verdere verdieping. Verslag van de vierde discussiebijeenkomst over biotechnologie bij dieren', Utrecht, Ethics Institute, 2004.

44 Bovenkerk and Poort, 'The Role of Ethics Committees'.

issues in the licensing procedure has become a kind of 'ritual dance'.[45] This 'ritual dance' refers to the repetitive nature of the interactions between the CAB and commentators on licensing applications. The dance starts with the tendency of a relatively small group of stakeholders (such as animal protection agencies) to submit the same objections over and over again. The CAB already anticipates these objections in their first recommendations. The stakeholders do not feel their views are taken seriously, and therefore submit them again.[46] This process has resulted in a standardization of both the stakeholders' objections and responses to these objections by the CAB. At the same time, despite slowing down the licensing procedure and failing to deepen understandings of the moral status of animals, these objections at least ensure that moral considerations are voiced in some way.[47]

Second, the stagnation of moral debate can be explained as a consequence of the legalistic context in which the debate takes place. The legalistic context of the licensing procedure has resulted in a process in which the scientists formulate their arguments with an eye to the legal language in which licensing decisions are made.[48] The legalistic context has arisen due to the CAB's obligation to defend their decisions before a legal review board, since both scientists and stakeholders have a right to appeal. Its reasoning takes place in the context of legal principles, norms, and values. It is therefore difficult for the CAB to revise their position or experiment with new ideas, since this could contradict the legal security of its final judgement.[49] The licensing procedure is thus in many ways no longer a moral reflection on the propriety of licensing, but a bureaucracy built on legal standards.

7.3 Characteristic Elements of a Legislative Approach

This section elaborates the three characteristic elements of the interactive legislative approach in light of the Dutch regulation of animal biotechnology: public involvement in legislation, the role of moral values in legislation, and legitimacy of legal norms.

7.3.1 *Public Involvement in Legislation*

The Netherlands considers public involvement valuable in developing norms on animal biotechnology. For example, one of the objectives of requiring ethical review of biotechnological procedures was to signal and provide opportunities for discussing problematic developments.[50] The review was meant to provide insights into the various viewpoints and

45 Paula, *Ethics Committees*, Bovenkerk and Poort, 'The Role of Ethics Committees'.
46 Paula, 'Biotechnologie bij dieren ethisch getoetst?'.
47 Interview with E. Schroten, Spring 2009.
48 Paula, 'Biotechnologie bij dieren ethisch getoetst?', Meijer et al., *Evaluatie van het Besluit Biotechnologie bij Dieren*.
49 See Paula, 'Biotechnologie bij dieren ethisch getoetst?'.
50 Explanatory Amendment of the Animal Biotechnology Decree; Paula, Biotechnologie bij dieren ethisch getoetst?; Meijer et al., *Evaluatie van het Besluit Biotechnologie bij Dieren*.

concerns about animal biotechnology in the pluralistic Dutch society. It was also intended to stimulate further societal discussion.[51] The Dutch government in general as well as the Minister of Agriculture in particular adopted a policy of stimulating the formation of public opinion. Brom has classified this policy as aimed at the further development of public morality.[52] However, it is also necessary to ask whether this strong focus on public opinion formation also ensures public participation.

In the legislative process leading up to the adoption of the Law and the Decree, the public played a prominent role, although not a central one. First of all, despite the attempt to stimulate public debate in 1993, the outcomes of the public NOTA meeting eventually were not reflected in the final regulations. Nevertheless, the meeting contributed to public debate: the exchange of knowledge and information during the debate contributed to opinion forming, and also stimulated further debate.[53]

Second, during the legislative procedure various organizations and institutions participated in similar debates and in decision-making in general. These organizations and institutions had various backgrounds. In addition to official political institutions, scientists, experts in the field of ethics, and interest groups were also involved. The Animal Protection Organisation's alternative draft regulation, for example, influenced current policy.[54] Furthermore, the media attention on the Gene-Pharming and IVO experiments led to public and political debate. The interplay between these debates ensured that public concerns were taken seriously. Public concerns were also voiced by the various interest groups that participated in the legislative process.

The licensing procedure acknowledges an official role for the public, and incorporates an official public consultation procedure in which the public can comment on the draft decision of the Minister and the CAB opinion.[55] In reality, however, the involvement and influence of the general public is limited. First of all, only a small group of stakeholders tends to submit objections.[56] Most of these stakeholders are opponents of the use of animal biotechnology and therefore do not represent the general public. Second, as mentioned in the previous section, these limited groups of stakeholders tend to submit standardized objections for which the CAB has already prepared standardized responses: the so-called ritual dance. Third, real debate during the consultation round is discouraged.[57] Consequently, public involvement seems to be a procedural necessity instead of an attempt to broaden viewpoints.

In response to this lack of meaningful public involvement, the CAB tried to break through this stalemate and initiate a series of broader debates on general themes concerning

51 Explanatory Amendment of the Animal Biotechnology Decree.

52 Brom, *Het toetsen van biotechnologische handelingen bij dieren*.

53 This already can be seen as a first tempt to develop public morality.

54 Parliamentary Documents, 166447, nr. 15-16.

55 Committee for Animal Biotechnology, 'Annual Report', Utrecht, Ministry of Agriculture, Nature Conservation and Fisheries, 1998.

56 Paula, 'Biotechnologie bij dieren ethisch getoetst?', Meijer et al., *Evaluatie van het Besluit Biotechnologie bij Dieren*.

57 Bovenkerk and Poort, 'The Role of Ethics Committees', p. 25.

animal biotechnology.[58] These meetings were partially successful, as they drew participation from new and different stakeholders, including proponents of animal biotechnology. At the same time, however, the meetings did not stimulate an ongoing learning process regarding animal biotechnology. No new arguments were brought to bear.[59]

7.3.2 *The Role of Moral Values in Legislation*

Moral values played a central role in the legislative process leading up to the adoption of the Law. Moral concerns about animal biotechnology strongly influenced the political agenda. In his dissertation, Brom argued that several arguments related to moral concerns influenced political decision-making on animal biotechnology issues. Concerns about animal integrity, playing God, naturalness, and the instrumentalist use of animals all had an impact on the legislative process.[60]

In the parliamentary debate, moral controversies were extensively discussed in the context of the intrinsic value of animals. Various institutions and committees were appointed to analyze the moral concerns about animal biotechnology and to develop a framework for ethical review.[61] The 1990 report of Schroten I presented recommendations for a framework for review that included reflection on natural values such as homeostasis, genetic biodiversity, the integrity of the individual, and the order of species. Furthermore, it suggested that decision makers should consider the intrinsic value of animals in terms of health and welfare, the health and welfare of human beings, protection of the environment, and ideological considerations in terms of creation and evolution.[62] The Minister of Agriculture has adopted a position that requires reflection on the intrinsic value of animals and the values of homeostasis and natural order of species.[63]

58 Visak and Meijboom, 'Integriteit van dieren: Bouwsteen of Struikelblok? Verslag van de 1ste discussiebijeenkomst over biotechnologie bij dieren', Visak and Meijboom, 'Op zoek naar alternatieven – maar voor welke doelstelling?'; Visak and Meijboom, 'Het maatschappelijk belang van een doelstelling: Zoeken naar goede argumenten. Verslag van de derde discussiebijeenkomst over biotechnologie bij dieren'; Visak and Meijboom, 'De waarde van een machtsvrije dialoog voor verdere verdieping. Verslag van de vierde discussiebijeenkomst over biotechnologie bij dieren'.

59 Visak and Meijboom, 'Integriteit van dieren: Bouwsteen of Struikelblok? Verslag van de 1ste discussiebijeenkomst over biotechnologie bij dieren', Visak and Meijboom, 'Op zoek naar alternatieven – maar voor welke doelstelling?'; Visak and Meijboom, 'Het maatschappelijk belang van een doelstelling: Zoeken naar goede argumenten. Verslag van de derde discussiebijeenkomst over biotechnologie bij dieren'; Visak and Meijboom, 'De waarde van een machtsvrije dialoog voor verdere verdieping. Verslag van de vierde discussiebijeenkomst over biotechnologie bij dieren'; Meijer et al., *Evaluatie van het Besluit Biotechnologie bij Dieren.*; Bovenkerk and Poort, 'The Role of Ethics Committees'.

60 Brom, *Onherstelbaar verbeterd.*

61 For example, Fretz and Sterrenburg, *Het maakbare dier geregeld*, which is an influential report for the moral reflection on animal biotechnology. In this report various arguments as well as various normative moral theories were analyzed.

62 Rapport van Tijdelijke Commissie van Ethiek en Biotechnologie bij Dieren 1990, NRLO/90/W 55, Den Haag.

63 Official Letter of the Minister of Agriculture, Nature and Fishery to the House Representatives, 1991, dd. 28 November 1991, 19744, nr. 8.

Finally, the Animal Health and Welfare Law recognizes the moral aspects of animal biotechnology and includes them in its section on licensing. According to the Law, a license can only be granted if no moral objections exist. The Explanatory Amendment discusses these moral objections in terms of health, welfare, and intrinsic value of animals.[64] However, this precondition for licensing has a primarily symbolic value, since moral objections will always exist. The regulation did not intend to establish a general ban on animal biotechnology, but rather to control its use. Some applications of the technology can be justified. Nevertheless, including rules regarding moral objections in the legal texts ensures that moral concerns are not entirely forgotten. The work of the CAB ensures a critical reflection on the moral acceptability of biotechnological procedures with animals.

In reality, however, the role of moral values is not that explicit. Despite the conditions ensuring critical moral reflection in the licensing procedure, moral reflection in licensing is in reality limited. The ritual dance and the legalistic context in which the CAB operates help to explain this.

First of all, the CAB must reason within the legalistic context of the licensing procedure. This legalistic context is problematic because it makes it difficult for the CAB to experiment with new ideas and to discuss controversial developments.[65] Second, the debate on the moral status of animals seems to be 'frozen', even though their status has been only partly clarified. There is no discussion anymore about which animals can be used and for what kinds of research.[66] The concept of intrinsic value has also not been further developed.

Paula has identified three different strategies that the CAB uses to develop decision-making norms.[67] First, the CAB has developed substantial consensus norms. Second, the CAB has developed various procedural norms for dealing with a lack of data or criteria. These procedural norms can postpone substantive judgement. Third, the CAB has developed a strategy of reframing the issue at hand with evasive argumentation, which Paula refers to as the semantic shift.[68] Paula argues that the last two strategies are intended to smooth over moral conflicts. In concrete cases, such a strategy may seem useful. However, in the larger sense these strategies will not contribute to long-run problem solving, as underlying issues continue to exist.[69] Paula therefore questions whether the CAB is an appropriate policy instrument for contributing to the deepening and clarification of these moral questions.[70] It is worth nothing here that the CAB is currently under revision, as the Minister of Agriculture argues that it has completed its task. Practically speaking, the CAB is no longer operative. This revision is remarkable because both the CAB's members and its critics point to the loose ends that still need to be tied. There is broad agreement that the work of the CAB is not done yet.[71]

---

64 Explanatory Amendment of the Animal Health and Welfare Law.

65 Bovenkerk and Poort, 'The Role of Ethics Committees'.

66 Interviews with E. Schroten and with S. Beukema (secretary of the CAB), 2009.

67 Paula, *Ethics Committees*, Ch. 3.

68 *Ibid.*, pp. 78-80.

69 Bovenkerk, *The Biotechnology Debate*, p. 178.

70 Paula, *Ethics Committees*, pp. 80-81.

71 Bovenkerk, *The Biotechnology Debate*, interview with E. Schroten, 2009.

In sum, while moral values played an important role throughout the formulation and implementation of Dutch animal biotechnology law, they have not been adequately developed in legislative practice. Instead of strengthening and clarifying the moral position of animals, moral reasoning in the licensing procedure has stagnated. The CAB's strategy and the Minister's approach have resulted in concrete standards in very few cases.

7.3.3 *Legitimacy of Legal Norms*

Like most western democratic societies, the Dutch consider that following official procedures will ensure the validity of legal norms.[72] As Eijlander et al. note, the Dutch legislature accepts other sources of legitimacy as well, such as legitimacy in terms of the people's acceptance of and willingness to follow legal norms.[73] Van Schooten explains the decreased focus on procedural legitimacy or validity as correlated with an increase in focus on establishing communicative legitimacy.[74] Van Klink distinguishes three requirements for legitimacy in the Dutch context: 1) equal representation; 2) decisions based on majority opinion; and 3) responsive and thorough reasoning in decision-making processes in which no one interest has absolute value. These requirements mainly refer to democratic legitimacy, rather than substantive legitimacy. Van Klink relates democratic legitimacy to a principle of consensus that is translated into a participation requirement.[75] Van der Vlies goes further, and holds up the principle of consensus as one of the principles of good regulation (*beginselen van behoorlijke regelgeving).*[76]

In short, the legitimacy of regulation in the Netherlands depends on validity, and most prominently on popular legitimacy in terms of acceptance, involvement, and consensus. However, as this section argues, for complex policy problems such as animal biotechnology, the legitimacy of legal norms requires an additional substantive justification. This section discusses both popular legitimacy and substantive justification as established in the Dutch regulations on animal biotechnology. It will start by elaborating the concept of popular legitimacy.

Popular legitimacy is mainly attained through broad deliberation and stakeholder participation. The legislative processes of the Law and the Decree featured broad deliberation and a search for consensus on the acceptability of animal biotechnology. In the Dutch context, consensus is an important element of establishing popular legitimacy.[77] Yet, no consensus existed with respect to animal biotechnology. Consensus could therefore not

72 Ph. Eijlander and W. Voermans, *Wetgevingsleer,* Den Haag, Boom Juridische Uitgevers, 2000, p. 25.

73 Eijlander et al., *Wetgevingsleer,* p. 26.

74 H. van Schooten, 'Van bevel naar communicatie', in B. van Klink and W. Witteveen (eds.), *de overtuigende wetgever',* Deventer, W.E.J. Tjeenk Willink, 2000, p. 73.

75 Van Klink, *De wet als symbool,* p. 148.

76 I.C. van der Vlies, *Het wetsbegrip en beginselen van behoorlijke wetgeving: het legaliteitsbeginsel,* Den Haag, VYGA, 1984.

77 Van Klink and van der Vlies refer to this as a principle for democratic legitimacy. See van Klink, *De wet als symbool,* p. 148.

serve as the foundation for legal norms. As a result, the legislature adopted open norms requiring further development.

Popular legitimacy also depends on public acceptance. The focus on public opinion formation and the development of public morality exemplify this interpretation of popular legitimacy. The Dutch legislature aimed to develop a more unified public morality concerning animal biotechnology.[78] In other words, one of the incentives for regulating animal biotechnology using this approach was to come to substantial consensus about the moral values and the legal norms involved. Popular legitimacy in this context depends on public debate, broad deliberation by experts and interest groups, and the provision of information.

Popular legitimacy is not enough, however: legal norms for the regulation of animal biotechnology also require substantive justification. The Dutch Minister of Agriculture adopted a 'no, unless' policy in protecting animals in the field of animal biotechnology. In principle, biotechnological procedures with animals are prohibited, unless important societal interests are at stake. This policy may appear at first sight to be substantive. However, as with the Danish principle of proportionality, it is actually more of a procedural justification, as it lacks content in itself. Decisions cannot be justified by the content of a 'no, unless' policy. However, this restrictive approach has a strong symbolic value.[79] By elaborating this policy, the Minister underlined the lack of consensus norms on public morality. The 'no, unless' approach attempted to overcome disagreement about moral considerations as well as to address the current need for regulation. Because moral concerns were the basis for adopting a restrictive approach, they provide some substantive justification for the policy and ensure the incorporation of some legal norms concerning moral reflection.[80]

Additionally, the reports of experts on the moral aspects of the regulation ensured that the policy, the use of open norms, the licensing procedure and its review framework were all justified in light of morality. The fact that the extent, the consequences, and the meaning of the animal biotechnology regulations were central to the political debate, also helped to establish a substantive justification for the legal framework. The 'no, unless' policy operated as a principle for ensuring a critical reflection on the morality of licensing applications. These aspects of the Dutch animal biotechnology regulations provided at least a partial substantive justification for the Law and the Decree.

7.4 Dutch Political Culture

This section discusses in three subsections the Dutch political culture. To begin with, Section 7.4.1 discusses the Dutch democratic traditions. Section 7.4.2 explores the role of public involvement in the broader political context in the Netherlands. This subsection also

78 Brom et al., *Het toetsen van biotechnologische handelingen bij dieren.*
79 Explanatory Memorandum of the Animal Health and Welfare Law.
80 Explanatory Memorandums of the Animal Health Welfare Law and the Animal Biotechnology Decree.

examines the role of expert involvement, which is useful for elaborating on the horizontality of decision-making processes. Finally, Section 7.4.3 examines the Dutch legislative culture.

7.4.1 Democratic Traditions

The Netherlands can be characterized as a consensual type of democracy[81] that is also termed a consociational democracy.[82] This section outlines the historical developments that led the Netherlands towards this type of democracy. A history of 'pillarization' (*verzuiling*) and a 'politics of accommodation' (*poldermodel*) have contributed to the establishment of a corporatist model in recent Dutch politics. The Dutch divide their recent political history into three periods of time: 1) 1917 through 1967, which was the period of pillarization; 2) 1967 through the late 1970s, which was marked by depillarization; and 3) the 1980s through today, which has been characterized by the '*polder model*'.[83]

Dutch democratic traditions are characterized as a type of consociational democracy.[84] The Dutch have a long history of compromise and tolerance in political decision-making. According to Lijphart, the constitutional amendment of 1917 marked a particularly important step towards a politics of accommodation, although the idea goes back much further.[85] Lijphart refers to this constitutional amendment as an important moment of change, since it represented an official step towards a politics of accommodation and consensual democracy. This amendment defused important political conflicts of the time, such as the school struggle and the emancipation of the electorate.[86] This accommodative politics is connected with pillarization, and can be explained as a politics of decision-making that operates by obtaining the cooperation of pillar elites.[87]

---

81 In a consensual type of democracy decisions are made based on a consensus between different minority groups. Although in the Dutch system it is aimed for consensus, it does not adopt a Habermasian interpretation of consensus. Instead its aims is to ensure that decisions can be agreed upon by those who have to work with it.

82 H. Bakvis, 'Towards a Political Economy of Consociationalism. A Commentary on Marxist Views of Pillarization in the Netherlands', in *Comparative Politics*, 16/3, 1984, p. 315.

83 See J. Stamhuis, *Conflicting Interests in Corporate Regulation*, Enschede, Gildeprint Drukkerijen, 2006, she refers to A. Lijphart, *The Politics of Accommodation: Pluralism and Democracy in the Netherlands*, Berkeley, University of California Press, 1975; T.R. Rochon, *The Netherlands. Negotiating Sovereignty in and Independent world*, Boulder Co, Westview Press, 1999; and S. Taekema, *Understanding Dutch Law*, Den Haag, Boom Juridische uitgevers, 2004.

84 The Netherlands is a constitutional monarchy with a parliamentary system which is recognized in its official procedures. The constitution of 1848 set the foundation for the parliamentary system of the Netherlands and the representative democracy. The Parliament, consisting of two Houses, is first and foremost co-legislator. The Second Chamber, directly chosen by the Dutch people, has political primacy in decision-making. The power of legislation is assigned to both the Parliament and the government. The government officially consists of the King and the ministers and also has the executive authority. However, in reality, the government initiates regulations for various topics.

85 Lijphart, *The Politics of Accommodation*, p. 27.

86 J. Den Hertog, "Stille Politiek', en de pacificatie van 1917', in D. Bos, M. Ebben, and H. te Velde (eds.), *Harmonie in Holland*, Amsterdam, Bert Bakker, 2007, pp. 176-193.

87 Lijphart, *The Politics of Accommodation*, p. 197.

It was not until after the Second World War, however, that pillarization and accommodation politics became truly ascendant. From the Second World War until 1967, pillarization and accommodation politics dominated political decision-making. the Netherlands was divided into four pillars: the Catholic pillar, the Protestant pillar, the Liberal pillar, and the Socialist pillar. These pillars each had their own political parties, newspapers, education, social networks and even trade union movements.[88] In particular, developments within the socialist movement towards a culture of corporatism increased the influence of accommodation politics.[89] According to Lijphart, the politics of accommodation seemed to function well because no one pillar included a majority of the Dutch electorate. Consequently, the Dutch government consisted of coalitions between the pillars. The politics of accommodation offered a pragmatic solution in which compromises between the pillars were reached by aiming for consensus.[90] The Dutch consensual style of democracy. was thus based on a corporatist decision-making model.[91] The elites played a key role in this politics, and their participation contributed to a high degree of political stability.

Furthermore, Lijphart states that this system functioned well in the Netherlands due to a rather passive population. It was argued that successful decision-making processes required meetings behind closed doors. By negotiating in private, the elites could discuss and come to an agreement on political issues without a loss of face towards their base. At the same time, the 'back-room politics' and bureaucracy were criticized for their lack of room for opposition, since this strategy reduced the need to justify decisions publicly.[92]

In the period from 1967 until the late 1970s, the stability of the political climate declined.[93] This period was marked by a process of ideological redefinition, depillarization, and polarisation. The socialists in particular objected to the system of pillarization and especially the role played by pillar elites. Citizens became more actively involved in politics, and demanded public justifications of legal decisions.

After a brief period of polarisation, these challenges for political parties came to an end. The fights had been fought, and from the late 1970s until today polarisation gradually disappeared.[94] The old '*polder model*' with its ideals of compromise and tolerance became popular once again. Although the term was only coined in 1998, the method dates back to the period of pillarization. The '*polder model*' is named after the reclaimed land the Netherlands is known for. This land can only stay dry if all parties involved are willing to co-operate on the basis of consensus.[95] As a form of political decision-making, it involves

88 Lijphart, *The Politics of Accommodation*, p. 23.

89 C. Jeurgens, 'Poldersporen uit het verleden: Een afspiegeling van ondoorzichtelijke besluitvorming', in D. Bos, M. Ebben, and H. te Velde (eds.), *Harmonie in Holland*, Amsterdam, Bert Bakker, 2007, pp. 216-235.

90 Lijphart, *The Politics of Accommodation*, p. 104.

91 A, Lijphart and M. Crepaz, 'Corporatism and Consensus Democracy in Eighteen Countries', in *British Journal of Political Science*, 21, 1991, pp. 235-246.

92 Bovenkerk, *The Biotechnology Debate*, p. 173.

93 Lijphart, *The Politics of Accommodation*, p. 197.

94 Stamhuis, *Conflicting Interests*, p. 50.

95 K. van der Bruggen, 'Dolly and Polly in the Polder: Debating the Dutch Debate on Cloning', in *The Public Debate on Cloning International Experiences*, Den Haag, Rathenau Institute, 1999, pp. 16-17.

cooperation between different interest groups that have a direct stake in the policy field. These actors reason together in a system of negotiation and aim for a consensus.[96]

Despite its international fame, the Dutch polder model has also been widely criticized. For example, Spoormans argues that this model implies not a culture of deliberation, but a strategy of avoiding conflicts.[97] This conclusion is reminiscent of Paula's argument that the CAB uses consensus and deferral as strategies to evade conflict regarding the moral questions brought up by animal biotechnology.[98] Furthermore, *'polder model'* politics does not involve a similar sort of exchange of arguments that is assumed in deliberative decision-making, but rather a process of negotiation of interests and beliefs.[99] In addition, this model is still characterized as elitist and non-transparent: the so-called 'back-room politics' continue.[100] Even if the pillar elites have now been partly replaced by other actors, the strategy of private, negotiated decision-making has not changed. To a certain extent, the consultation of experts from various fields ensures a kind of objectivity in decision-making that counterbalances the power of interest groups. At the same time, however, disagreements among interest groups and their splintered backing make decision-making processes seem even less transparent.[101]

In short, Dutch political culture, which has been characterized by pillarization, de-pillarization, and a politics of accommodation, has resulted in a system of corporatist decision-making, in which co-operation between various interest groups is central. Nowadays, the pillars have almost disappeared, but old consociational decision-making strategies are still used. Despite its 'back-room politics' and the high levels of bureaucracy, corporatist decision-making operates horizontally. However, this decision-making strategy does not guarantee that decisions are actually based on broad deliberation.[102]

7.4.2 The Role of the Public

Section 7.3.1 outlined the role of the public in the legislative process of the Law and the Decree. During the legislative procedure, the public's official role was limited, although experts and interest groups were heavily involved. In addition, the need for public debate and public opinion formation were explicitly emphasized. In the licensing procedure, the public was authorized to object to the applications. This ensured a moment of public input by creating possibilities for objecting to licensing applications in a public hearing. In reality, however, the influence of the public on decision-making was limited.

96 J. Stamhuis, 'Communicative Law: A Quest for Consensus', in N. Zeegers, W. Witteveen, and B. van Klink (eds.), *Social and Symbolic Effects of Legislation under the Rule of Law*, Lampeter, Ceredigion, Wales, The Edwin Mellen Press, 2005, p. 226.

97 Den Hertog, 'Stille politiek', pp. 176-193.

98 Paula, 'Biotechnologie bij dieren ethisch getoetst?'.

99 Bovenkerk, *The Biotechnology Debate*, p. 172.

100 R.B. Andeweg and G.A. Irwin, *Governance and Politics of the Netherlands*, Basingstoke, Palgrave MacMillan, 2002, p. 42.

101 See Stamhuis, *Conflicting Interests*, pp. 52-53.

102 See Paula, 'Biotechnologie bij dieren ethisch getoetst?'; and Meijer et al., *Evaluatie van het Besluit Biotechnologie bij Dieren.*

This section highlights the role of the public in the broader Dutch political culture. The public role has changed over the years with the pillarization and depillarization of current politics, but has traditionally been characterized by representation. During the period of pillarization, the public remained passive. Decision-making was left to the elites, who represented their pillar in all kinds of fields. These elites explicitly identified with their grassroots. Neither the public nor the elites saw any use for further public involvement.

Even after depillarization, traces of the old system can still be recognized. the Netherlands is known for its consensus-oriented decision-making culture. that was traditionally linked to pillarization. The pillar elites have now been replaced by other actors. Citizens are no longer represented by one pillar elite for all kinds of different subjects, but by representatives of organized stakeholder groups that actively participate in decision-making processes.[103] The public's interests and preferences have become more splintered. To be sure that all of their individual interests are represented, citizens must have a more active attitude. They can no longer trust one single actor to represent all their interests. Consequently, decision-making has become broader and more horizontal.

Furthermore, interest groups are less dependent on their grassroots than were the pillar elites, which makes representation less direct. This seems to be one of the pitfalls of the new polder model. To what extent are the citizens actually represented and involved?[104] Chapter 4 distinguished two conceptions of the public: the public as one actor and the public as a collection of interests. The '*polder model*' reflects the latter idea, since the public as one actor does not have a place in the Dutch political process.

The Dutch government has attempted to overcome these pitfalls by creating various possibilities for public involvement in the Dutch legal system. For example, the government initiated joint decision-making procedures at a decentralized level and institutionalized an optional referendum. Especially at the local level, decision-making processes are open to the public, which can lodge objections to regulations or ongoing discussions.[105] Furthermore, the public has similar possibilities in all kinds of licensing procedures.[106] These official procedural rules improve democratic legitimacy and increase the transparency of decision-making. Especially at the decentralized local level, the public can be and does feel involved.[107] These possibilities for public input do not necessarily contribute to well-considered decisions, however. In reality, the public's input has often been channelled into bureaucratic standardized procedures.

Direct public participation at the national level is limited. The public's official role in law making is restricted to the optional referendum and parliamentary elections in case of constitutional referenda, which in reality are seldom called for. Instead, public input depends on representation via organized interest groups and expert involvement.

---

103 Lijphart, *The Politics of Accommodation*, pp. 196-201, Bakvis, 'Towards a Political Economy of Consociationalism'.

104 Stamhuis, *Conflicting Interests*, pp. 52-53.

105 Algemene Wet Bestuursrecht.

106 For example, the licensing procedure for the Animal Biotechnology Decree.

107 F. Hendriks, *Vitale democratie, theorie van democratie in actie*, Amsterdam, Amsterdam University Press, 2006.

The experts often have no explicit role in representing the public, but instead give advice to politicians and legislatures. Even with a broad and horizontal process, this decision-making strategy still runs the risk of 'back-room politics'.[108]

Nevertheless, the feedback from public debate helps to establish transparency in decision-making. The stimulation of public debate and public opinion formation have a prominent position on the political agenda. On animal cloning and other controversial issues like euthanasia, abortion, gay rights, freedom of speech, and education, public debate and political debate influence and reinforce one another. There exists a feeling that only by discussing controversial matters in the open is it possible to develop adequate regulations.[109] Discussing everything is part of the Dutch culture. Kennedy, for example, has argued that in the context of euthanasia the tendency to make things easy to discuss has opened up possibilities for a progressive policy with clear regulations.[110]

7.4.3 *The Legislative Culture*

A country's legislative culture constitutes its perceptions of law and legislative approaches, and additionally affects ideas about the legitimacy of legal norms and the functions of law. Legislative culture is influenced by democratic and legal traditions. Section 7.4.1 discussed the Dutch democratic traditions. This section explores Dutch legal traditions.

The legal traditions of Western democratic countries are subdivided into several legal families. The two main legal families are common law and civil law. The legal systems in continental Europe mainly belong to the latter family. The civil law tradition is characterized by abstract reasoning and an emphasis on statutory law. The Dutch legal tradition is no exception to this. More specifically, the Dutch codes are modelled on French codes, and therefore the Dutch legal tradition has been classified as Romanistic. Various Romanistic institutions, for example the judicial structure, can be distinguished in the Dutch legal system.

The Dutch system is not purely Romanistic, but has been influenced by other legal systems as well. The most significant way in which the Dutch system differs from a strictly Romanistic one concerns the recodification of civil law. Here the Dutch system draws heavily on the German Civil Code. Furthermore, the Dutch Civil Code also incorporates some typically Dutch elements based on Dutch case law.[111] In particular, Dutch public law differs from the traditional concept of public law as mainly consisting of clear concrete norms, and adopts principles reminiscent of civil law, which reflects a more open interpretive framework.[112]

Several different schools of thought in legal philosophy and legal theory have influenced the Dutch legal system. To begin with, Scholten has been particularly influential

108 Bovenkerk, *The Biotechnology Debate*, p. 173.

109 J. Kennedy, *Een weloverwogen dood. Euthanasie in Nederland*, Amsterdam, Prometheus, 2002, pp. 18-19.

110 *Ibid.*

111 R. Lesaffer, 'A Short Legal History of the Netherlands', in S. Taekema (ed.), *Understanding Dutch Law*, Den Haag, Boom Juridische uitgevers, 2004, pp. 31-58.

112 Eijlander et al., *Wetgevingsleer*, Ch. 2.

in private law.[113] His work discussed an interpretive style of judicial decision-making in which judges partly base their decisions on moral intuition. These intuitions are developed through experience in legal decision-making.

Some argue that the Dutch school of thought in legal theory is a middle course between the traditional schools of natural law and positivism. Others see the system as instrumentalist. However, in legislative practice interactionism has become most prominent in Dutch legal thought. In interactionist theory, general legal principles function as a normative basis for law that is concretized in legislative practice.[114] The interactionist school of thought, influenced by Fuller, emphasises the ways in which norm development depends on the interplay between facts and values. It argues that law is identified with concretising ideals and the principle of law in action. Legal norms are developed by actions based on a shared interpretation of values.[115] These ideas can be recognized in regulating animal biotechnology, in which moral values played a prominent role.

The movement toward a more interactionist perspective was a response to the regulatory crisis in the eighties. During this regulatory crisis, two main problems appeared: over-regulation and qualitatively poor regulation.[116] A solution was sought, on the one hand, in developing more general legal principles and quality criteria for regulation,[117] and, on the other, in a more interactionist view of legislation.[118] Van der Vlies, for example, developed ten general principles to which regulation should respond.[119] Both of these strategies exemplify movement towards interactionism in the Dutch legal system.

This shift towards an interactionist view of law has not only led to the inclusion of principles and ideals in legal reasoning, but has also resulted in a process of rule-making and decision-making based on negotiations between various organized stakeholders and politicians with a focus on consensus and compromise.[120] Interaction among stakeholders is required in order to establish democratic legitimacy and develop norms with broad public support. No party has the prerogative of making decisions single-handedly. In other words, no one interest can overrule any other. Therefore, decisions should always be reconsidered and challenged.[121] Consequently, facts and norms are strongly related in the Dutch legal context.

These developments in legal thought have also influenced the Dutch legislative approach. The Dutch legal system acknowledges various legislative approaches for addressing diverse kinds of policy problems. Due to the influence of consensual democratic.

113 N.F. Van Manen, 'Rechterlijke rechtsvinding en intuïtie', in E.A. Lissenberg, P.W. Brouwer, F. Jacobs, and E. Poortinga (eds.), *De actualiteit van Paul Scholten,* Nijmegen, Ars Aequi Libri, 1996, pp. 57-74; Asser-Scholten, *Algemeen Deel,* Zwolle, W.E.J. Tjeenk Willink, 1974 (3[e] druk).

114 B. van Klink, 'Legislation in the Netherlands: The Communicative Turn', in S. Taekema (ed.), *Understanding Dutch Law,* Den Haag, Boom Juridische uitgevers, 2004, pp. 103-129.

115 Fuller, *The Morality of Law.*

116 Witteveen, 'Turning to Communication in the Study of Law', p. 19.

117 Nota Zicht op Wetgeving, WODC, 1991.

118 Witteveen, 'Turning to Communication in the Study of Law', p. 19.

119 Van der Vlies, *Het wetsbegrip en beginselen van behoorlijke wetgeving: het legaliteitsbeginsel.*

120 Van Klink, 'Legislation in the Netherlands: The Communicative Turn'.

121 Bovenkerk and Poort, 'The Role of Ethics Committees'.

traditions, the acknowledgement of various general legal principles, and the influence of interactionism, the Dutch legislative approach is pragmatic and flexible.[122] In the Netherlands, legislation can have communicative and symbolic functions in addition to the traditional constitutional and instrumental ones.[123]

The multiple functions of Dutch regulation require different legislative approaches depending on the issue addressed. These legislative approaches involve a variety of legal frameworks. Legal frameworks with clear directed norms relate to traditional instrumental functions, whereas more open legal frameworks aim to establish communication. Furthermore, the variety of legislative approaches also imply their own strategies for norm development. For example, negotiated decision-making involves broad participation in norm development.[124] In her dissertation, Stout notes a gap between the sociological and normative realities of the law. According to her, this gap can only be bridged by interaction with the actors involved.[125]

Indeed, it has been suggested that a more communicative legislative approach has replaced the traditional instrumental legislative approach in Dutch legal culture.[126] Dorbeck-Jung, for example, has suggested that the Netherlands has moved toward a more responsive legislative theory.[127] This move brings with it a new preference for interactive and communicative approaches.

7.5 CONCLUSION

In the Netherlands, most of the tenets of the interactive legislative approach are recognizable in the legislation on animal biotechnology. This outcome is not really surprising since the reconstruction of the interactive model of law is based on the ideas of various Dutch legal theorists. Their claims are partly based on empirical findings, which gives insight into the fact that the Dutch political culture demonstrates sufficient conditions for the application of an interactive legislative approach. Consensualist democratic. traditions have resulted in a decision-making strategy characterized by negotiation among various organized interest groups and a search for consensus: the polder model. Experts are often consulted during decision-making processes, which has created a horizontal structure.

122 S. Taekema, 'Introducing Dutch Law', in S. Taekema (ed.), *Understanding Dutch Law*, Den Haag, Boom Juridische uitgevers, 2004, pp. 17-30.

123 Eijlander et al., *Wetgevingsleer*, Ch. 2.

124 See for example N. Florijn, 'Open wetgevingsproces, communicatieve wet? De totstandkoming van de Electriciteitswet 1998', in B. van Klink and W. Witteveen (eds.), *De overtuigende wetgever*, Deventer, W.E.J. Tjeenk Willink, 2000, pp. 211-224; and M. Hertogh, 'Deal and Dilemmas in Dutch Governance: Negotiated Decision-Making in the Netherlands', in B. De Waard (ed.), *Negotiated Decision-Making*, Den Haag, Boom Juridische Uitgevers, 2000, pp. 11-24.

125 H.D. Stout, *De betekenissen van de wet. Theoretisch-kritische beschouwingen over het principe van wetmatigheid van bestuur*, Zwolle, W.E.J. Tjeenk Willink, 1994.

126 Van Klink, 'Legislation in the Netherlands: The Communicative Turn'.

127 B.R. Dorbeck-Jung, 'Een theoretisch kader voor wetgevingstoetsing. Op weg naar een responsieve wetgevingsleer', in *RegelMaat*, 1998/4, pp. 198-209. Dorbeck-Jung refers here to Perelman, a Belgium legal theorist whose ideas have had prominent influence in the Netherlands.

Concerns, preferences, and interests from various perspectives are included in the legislative process. Furthermore, due to the influence of interactionism and a switch from traditional to more communicative legislative approaches, the decision-making process is open to input from various perspectives. The pragmatic attitude of the Dutch underlines the focus on an efficient solution rather than on enforcement of the rules. This attitude involves flexibility in legislative approaches, for example by recognizing the need for further norm development.

The elements of an ideal-typical interactive legislative approach are identifiable in the Dutch legislative approach to regulating animal biotechnology, as well as in Dutch political culture in general. The involvement of experts and interest groups characterizes the decision-making process. In the legislative process, public debate played an important role in setting the political agenda. The process of norm development was quite interactive.

The licensing process also features public involvement. Expert groups like the CAB as well as the public in general have a say in the licensing process, and public debate is explicitly stimulated. However, open processes and official possibilities for broad participation are no guarantee that communication and deliberation will occur in legislative practice. As we have seen, the Dutch licensing procedure ended in a ritual dance in which the same objections were expressed over and over again. The CAB anticipated these objections and standardized its rejections. Input and participation were transformed from an opportunity for broadening and deepening input into considerations about licensing to a mere bureaucratic necessity.

The Dutch legislative approach also has elements of a dynamic process. The regulation of animal biotechnology included various open aspirational norms that required further development. Further norm development was an explicit objective of the Dutch legislature. The licensing procedure, the functions of the CAB, and the input of stakeholders in licensing all seem to contribute to a dynamic process of norm development. A learning process concerning the moral status of animals has been at least partially established. Standards concerning the moral acceptability of some biotechnological procedures involving animals have been developed. Consensus has been reached on the use of genetically altered mice and rats for research into life-threatening diseases.

However, the learning process has not reached a higher aggregation level. The process of norm development has stagnated. Moral questions were 'frozen' and therefore the moral status of animals has not been further clarified or strengthened. The CAB's role in clarifying and strengthening the moral status of animals was limited by the legalistic context in which it took place. The CAB had difficulties experimenting with new ideas and changing its final decisions. Additionally, existing standards were sufficient to deal with most of the licensing applications, and therefore did not require further debate. A dynamic norm development process was, consequently, only partly established. Therefore, the Dutch regulation of animal biotechnology does not conform to all parts of the ideal-type of the interactive legislative approach.

The Dutch case study gives rise to questions concerning the role of ethics committees, the possibilities for a dynamic process of norm development, and the adequacy of the interactive legislative approach. The next chapter will compare the outcomes of the three case studies and discuss the questions raised during this analyzes.

8 Reflections on the Case Studies

8.1 Introduction

This chapter presents a comparative analysis of the Danish, Swiss, and Dutch case studies in light of the ideal-typical model of the interactive legislative approach. The analyses of these case studies in the previous three chapters showed various discrepancies and tensions between the theoretical background of the ideal-typical interactive approach and the legal and political reality in each country. This chapter presents the outcomes of these case studies and examines them alongside the ideal-typical model developed in Chapter 3.

The ideal-typical model consists of various theses, assumptions, and hypotheses. In order to examine the interactive legislative approach's practical value, the case studies' outcomes will be confronted with the subset of assumptions and hypotheses used in Chapter 4 to structure the case studies. This chapter analyzes whether these could be corroborated in the case studies. Furthermore, it will make some evaluative remarks about the practical value of the interactive legislative approach in Denmark, Switzerland, and the Netherlands. I do not aim to evaluate the quality of the legislative approaches used in these countries. Instead, the evaluation presented here will raise questions about the assumptions and hypotheses, and assess their usefulness for achieving the objectives of the interactive legislative approach: 'interaction' and 'dynamics'. Further reflection on the adequacy of the interactive legislative approach and the ideal-typical model will be reserved for Part III, which will attempt to provide some explanations for the shortcomings of the dynamic process and give some suggestions for changing and refining the ideal-typical model.

This chapter will be organized as follows: Section 8.2 will discuss the outcomes of the case studies in light of the seven assumptions and hypotheses of the ideal-typical interactive legislative approach. Section 8.3 will elaborate on the conditions necessary for an interactive legislative approach to function well. It will focus on the democratic traditions as well as the legal culture of each of the countries studied. This section also discusses the role of ethics committees. Finally, Section 8.4 will draw some conclusions about the adequacy of the interactive legislative approach and will point out the difficulties faced by a dynamic process of norm development. These difficulties will be addressed in Part III.

8.2 Characteristic Elements of the Interactive Legislative Approach

8.2.1 Co-Operative Effort

Various social actors, including governmental bodies and citizens, are involved in a co-operative effort on a horizontal level during all phases of norm development.

In general, the processes for regulating animal biotechnology in all three countries corroborated this claim. During the legislative process as well as the implementation phase in all three countries, both norm development and decision-making with respect to animal biotechnology regulations involved broad and horizontal processes. These processes were to a certain extent based on a co-operative effort. However, the role of the public was limited and indirect, often operating via representatives.

Denmark, for example, set up a committee to analyze the regulatory possibilities for animal cloning. Chapter 5 describes this committee that consists of experts from various fields, including governmental bodies. This preparatory committee presented a report that included an overview of the most prominent reasoning and suggestions for regulating animal cloning and genetic modification. This report was the point of departure for drafting the final law on animal biotechnology. Section 5.2.1 and Section 5.4.2 described some of the Danish extensive participatory methods it is known for as well as its focus on expert involvement in lawmaking. It solicits public and expert opinions through methods such as consensus conferences, citizen juries and expert panels.

At the same time, Chapter 5 criticized the consensus conferences, as they seemed to include in their final reports only those viewpoints that were consented to by all lay members. The reports excluded any viewpoints that were not shared by everyone, even if they represented a significant share of public opinion. Agenda-setting, policies, and decisions built on these final reports, therefore, are only practical products of public involvement. Nevertheless, due to such methods the Danish people feel involved in the legislative process.[1]

In Switzerland, the regulation of animal biotechnology was a response to an earlier amendment of the Swiss Constitution in which the Swiss People voted to protect the dignity of living beings. As discussed in Section 6.2.1, the legislative process leading up to the adoption of the Gene Technology Law (GTL) included a pre-parliamentary consultation procedure in which various experts and social actors could express their concerns, opinions, and interests. This procedure had a major influence on the final legislation, as the legislature considered all comments and included the most relevant ones in its draft.

As in Denmark and Switzerland, the Dutch legislative process was also not restricted to the parliamentary arena. Various interest groups and research groups were consulted and involved in developing the regulatory framework. For example, Section 7.2.1 discussed how an alternative proposal for the Animal Health Act drafted by the Animal Protection Organisation influenced discussions about the policy for protecting animals. Although the Animal Protection Organisation's proposal was not accepted by Parliament, their ideas for a stricter animal protection policy coloured parliamentary debate. Furthermore, ethics expert committees provided moral reflection on animal welfare and value issues.

In all three countries, the regulations on animal biotechnology prescribe a licensing procedure. The implementation phase of these regulations therefore runs parallel with the licensing procedure. These licensing procedures, like the legislative procedures that created them, are also structured by broad decision-making processes.

1 Section 5.4.2.

Section 5.2.2 discussed the Danish licensing procedure. Although the Danish legislature did not intend to stimulate ongoing norm development, the licensing process is a deliberative process that includes a variety of actors. The responsible authority is a rather broad assembly of experts of various fields and interests. Therefore, licensing decisions are based on a co-operative effort among various actors. However, in light of the interactive legislative approach, the licensing process still leaves much to be desired, as it comes to providing a forum for public debate and for public involvement. For example, during licensing, the public has no possibility to express concerns or object to decisions. Furthermore, the licensing procedure has become standardized to the point that there is no room for substantive deliberation among the members.

In Switzerland, the governmental body that is responsible for licensing is obliged to consult various expert committees and other governmental bodies. However, decision-making in the Swiss licensing procedure is less horizontal than in Denmark, as the scope of actors is limited.[2] Neither the public as such nor organized interest groups can participate in the licensing procedure. Expert committees are consulted for advice on licensing. However, their statements represent a single perspective on licensing and do not involve comprehensive advice or deliberation. For example, they are often limited to discussions of risk or moral reflections. At the same time, this expert input ensures that licensing is discussed from various viewpoints.

In the Netherlands, licensing is officially directed by the Minister of Agriculture. An ethics committee (CAB) advises the Minister on licensing. In the licensing procedure, however, the CAB has the leading role. At first sight, licensing seems to involve a horizontal decision-making process: the CAB co-operates with the Minister in decision-making. Furthermore, the licensing procedure includes a public hearing in which the public can object and comment on licensing applications. However, the Dutch licensing procedure is less horizontal as it seems at first sight, as in reality it is the CAB that has an authority to decide on licensing. Most remarkable is that the CAB also chairs the public hearings. Additionally, despite the room for public input, public objections are not directly taken into consideration by the Minister. Instead, it is the CAB that responds to the objections of the public and interest groups. In relation to the public, therefore, the CAB still occupies an authoritative position. Moreover, the CAB no longer adequately evaluates these comments, as over time it has developed formulaic standards for addressing public objections – which are, themselves, now often built on formulaic standards. Consequently, these rounds of objections no longer influence decision-making, but instead are merely a requirement prescribed by the law.[3]

Chapter 4 presented two concepts of the public. First of all, the public can be seen as a single actor, as being a collective identity. It then legitimates the whole political system. Second, the public can be considered a group of various actors. Then, the political influence that the people have, is a participatory one. As a single actor, the public was in the

2 Section 6.2.2.
3 Section 7.2.2 and Section 7.3.1.

three case studies primarily involved in decision-making via public debate. In the Netherlands, additionally, draft decisions for licensing are open for public objections. Furthermore, in all three countries the public was represented in decision-making by various groups of actors. In the Danish context, public involvement is generally stimulated by a variety of policy instruments that build bridges between scientists, citizens and governments. In Switzerland, various splintered interest groups can and did express their concerns and ideas explicitly in the pre-parliamentary procedure of the legislative process.

However, public involvement was limited. In Danish context, well-known methods for participation are not very prominent in the norm development process concerning animal biotechnology. In Switzerland, the public as well as various splintered interest groups are not well-represented in the licensing procedure. In the Netherlands, the rounds of public hearings are in reality a bureaucratic necessity instead of an instrument for ensuring public input.

8.2.2 Use of Open Norms

Open, aspirational norms play an important role in structuring and guiding the ongoing process of norm development. By the same token, an ongoing process of norm development guides the modelling of open, aspirational norms.

This hypothesis was not corroborated in the Danish case, and only partially corroborated in the Dutch and Swiss cases. The Danish legislature did not incorporate open norms in the regulation of animal biotechnology. Only during the legislative process there was room for discussing dignity of animals as an open, aspirational norm that needed to be taken into account for regulating animal biotechnology. The Dutch and Swiss legislatures attempted to stimulate an ongoing process of norm development by including open norms in the regulatory framework. However, in reality, an ongoing process of norm development has not been or has been only partly established.

The Danish legislature did not aim to establish an ongoing process of norm development when regulating animal biotechnology. Section 5.4.2 explained the lack of open norms in the regulatory framework by reference to the Ministry of Justice's strong instrumentalist view of law. The Ministry of Justice saw no need to respond flexibly to future trends in biotechnology. In general the Danish Minister of Justice does not acknowledge either the communicative or expressive functions of law, and consequently in this case rejected vague and open concepts such as 'dignity of animals'. Instead, the Danish legislature determined that a balancing of pros and cons, including moral values such as the dignity of animals, is sufficient for dealing with controversial issues.[4] Further norm development in the legal context is not required within the framework of the law.

In the Netherlands and Switzerland the legislature used open norms to leave room for further development in the field and responses to new trends. Moral values are included

4 Section 5.2.1 and Section 5.2.2.

in the legislation on animal biotechnology as aspirations for guiding further development of the legal norms. Although it was not exactly clear how the moral values should be interpreted, it was underlined that violation of these concepts would lead to various moral objections. The establishment of a licensing procedure brought further norm development into practice. In both countries, moral objections have to be taken into account in evaluating each licensing application. Expert committees play a prominent role in this process.[5] In the Netherlands, it was argued that this approach would lead to both development of the legal norms and the further crystallization of moral norms and values.

However, as previously noted, norm development has stagnated in both countries. In Switzerland, the concept of 'dignity of living beings', which is included as an open norm for protecting the moral status of animals, has been only partly defined in licensing procedure. The legal norms have been further developed in the sense that the licensing committee has established standards for practical cases. But more substantial definitional work has not occurred. During the legislative process, the Ethics Committee on Gene Technology in the Non-Human Field (ECNH) noted that the 'dignity of living beings' was a gradual concept and emphasized the need for further elaboration. The responsible authority applies this interpretation, but did and does not stimulate or elaborate any further norm development concerning the moral status of animals or the dignity of living beings.[6]

In the Netherlands, the intrinsic value of animals was both an aspirational norm that guided and structured further legal norm development, as well as an evolving norm, itself in need of development to clarify and strengthen the moral status of animals. The intrinsic value of animals as such is not included in the legal framework, but at least in the explanatory memorandum that in the Netherlands belongs to the legal texts. However, in reality, further development has not taken place. Despite the attempts of the responsible authorities and the CAB to break through the stalemate of repeating formulaic arguments, norm development is caught in a ritual dance.[7] Norm development concerning animal integrity has not reached a higher aggregation level and was limited to standardizing the results of the public hearings as well as the licensing procedure. The anticipated mutual-reinforcement of development of legal and moral norms has not taken place in legislative practice.

8.2.3 Moral Values

Moral values related to the controversial policy issues at stake are incorporated into all stages of norm development.

This claim interacts with the previous one, since the need for open norms and an ongoing process of norm development is primarily inspired by the moral values at stake. Moral values play a twofold role. First of all, moral values can function as a basis for moral reflection

5 Section 6.2.2 and Section 7.2.2.
6 Section 6.3.2.
7 Section 7.2.2 and Section 7.3.2.

in the legal context. Second, moral values represent a set of values that require consideration in themselves. The Danish legislature incorporated moral values in the legislative process, but did not attempt to institute an open process of norm development or take morality into account in the regulatory framework or implementation phase. The Swiss and Dutch legislatures attempted to create an open process in which moral values could play a role in the development of legal norms. Moreover, the Dutch legislature aimed to deepen and strengthen the moral status of animals. This attempt, however, was only partially successful. Rather than being integrated throughout the Dutch legislative process, moral questions were segregated in the licensing procedure.

In Denmark, moral values had a role in the legislative process, but were less prominent in the regulatory framework or in the implementation phase. Moral values were considered to be one of the 'interests at stake' in biotechnological procedures with animals, and as such were included in the development of legal norms. Legal norms were developed based on a principle of proportionality, which involves a balancing of pros and cons. For this balancing, moral values were translated into 'interests' that needed to be taken into account. But as 'interests', these values had a merely political meaning. This translation involves a downplaying of the values in an early phase as these cannot reflect and challenge an issue's moral impact in legal decision-making procedures. It is no longer possible to reflect on further development or new insights in the moral situation. In addition, the potential moral issues that may rise by further technological developments and developments in society can no longer be incorporated into the legislative procedure or, consequently, the regulatory framework it produced.

Moral values are not included in the legal texts or in the implementation phase.[8] Instead, it is argued that moral values has already been dealt with in the balancing of pros and cons that took place during the legislative procedure. There is therefore no need to express these values explicitly in the legal framework, or for further debate or interpretation of these values in the legal context.

In Switzerland, by contrast, moral values played a role in all stages of norm development in the regulation on animal biotechnology. First of all, moral values were discussed and incorporated into the legislative process. Second, the 'dignity of living beings' was included in the regulatory framework. And third, due to their status in the regulatory framework, moral values are open for further construction and interpretation during the implementation phase.

In reality, however, reflection on, construction of, and interpretation of moral values in the implementation phase is limited.[9] First of all, moral reflection takes place only in those cases that may lead to moral controversies. This limitation is not necessarily a disadvantage. Because the ECNH is not obliged to advise on each application, its role is really one of advising on only the moral controversies. The ECNH is thus not restricted to a legal analysis, but can concentrate on moral reasoning.

8 Section 5.3.2.

9 Section 6.3.2.

Second, the input for constructing and interpreting moral values is mainly restricted to the contributions of the ECNH. Concerns and comments of the public are rarely included. This would not be problematic were it not for the fact that the ECNH is supposed to stimulate public debate. Due to the ECHN's supposed role in public debate, the public is not represented via any other channel. In reality, though, the ECNH does not represent the public, and public opinion is, thus, excluded from licensing.[10]

Third, the licensing committee's interpretation of moral values is limited by earlier interpretations of the ECNH. Again, this is not problematic in itself. This approach promotes the development of the 'dignity of living beings' concept into standards that can easily be applied in the licensing procedure. However, focusing solely on the translation of the 'dignity of living beings' into practical standards leaves less room for questioning the fundamental basis of the concept itself. For example, in the ECNH's initial statement, dignity is explained as a gradual concept. But defining dignity as a gradual concept was not consented to by all actors, and required further debate. These basic questions are not addressed in current licensing procedure, which consequently limits the development of the concept at a higher aggregation level.

In the Netherlands, too, moral values have had a place in all three stages of norm development.[11] Moral values were widely discussed during the legislative process, and have influenced legal norm development. During the legislative process, legal and moral norm development strongly intertwined. In addition, the regulatory framework prohibits the granting of licenses where moral objections exist. The Explanatory Memorandum for the regulation on animal biotechnology articulates these moral objections in terms of animal integrity. Furthermore, due to the prominent role of the CAB and the provisions about moral objections in the regulatory framework, moral reflection takes place in licensing as well. Nevertheless, moral reflection is undermined by the legal context in which the CAB has to reason. Consequently, the moral impact is side-tracked by legal reasoning. The legal context in which the CAB has to operate seems to dominate.

8.2.4 Consensus

(Moral) consensus is not presupposed as a starting-point for norm development, but is instead regarded as one of the purposes of norm development.

In all three countries, consensus played an important role in norm development. Moreover, in all three countries, consensus is regarded as one of the purposes of norm development. The search for consensus structures decision-making and norm development.

In Denmark, the search for consensus played an important role in the norm development process. However, this differed from the assumptions about consensus in the ideal-typical model of the interactive approach. In the ideal-typical model, the norm

10 Section 6.3.1.

11 Section 7.3.2.

development process continues after implementation of the law, and the search for consensus justifies and structures this continuous process. Consensus is not required at the moment of constitution, but is, instead, a regulative purpose. In Denmark, by contrast, norm development on animal biotechnology regulations was restricted to the legislative process. Norm development ended with the constitution of the law.

Furthermore, a well-known method for public participation in Denmark are the consensus conferences. As mentioned before, lay panels discuss issues in organized conferences during which they have the possibility to ask questions to the experts. In the end, the lay panel present a report which includes the issues for which they have reached consensus. Purposes of these conferences are to increase public support and to improve deliberation. In reality, the influence of these reports is doubtful. Among other reasons, it can be questioned whether these reports can represent the public opinion adequately. Considerations that lead to major or even small controversies are not included in these reports, while those consideration are also part of the public opinion.[12]

In addition, the Animal Ethics Committee (AEC) aims for consensus in its deliberations and presents its statements as based on consensus. As a result, in the case study, the Minister of Justice could easily adopt their statements, arguing that they did not require further extensive deliberation in the Ministry due to their consensus-based status.[13]

In Switzerland, consensus was considered an end, but also a starting point for norm development. At the beginning of the process, there was no consensus on the concept of dignity of living beings.[14] Actors consented on the importance of the concept, but not on its meaning. The ECNH emphasized the need for further elaboration, as it made clear that not all members consented on the preliminary conceptualisations. In that sense, consensus was considered an end that still needed to be sought. A search for consensus was, thus, supposed to structure norm development.

At the same time, the gradual conception of dignity of living beings was presented as based on a consensus, and it functioned as a basis for the regulatory framework. Consequently, consensus also functioned as a starting point for further norm development in the regulatory framework.

Nevertheless, in reality, further debate has not been established. In the ideal-typical model, consensus was conceived as a means of structuring debate and ensuring open attitudes, as it was designed to stimulate a constructive interaction. However, despite the emphasis on consensus as a motivation for further debate, the responsible authorities adopted and applied the ECNH's interpretation of the concept and did not further question its meaning.[15] Rather than establishing a dialectic, the connection between consensus as a starting-point and consensus as an end of the norm development process seem to have counteracted further development.

12 Section 5.4.2.
13 Section 5.4.2.
14 Section 6.2.1.
15 Section 6.3.2 and Section 6.3.3.

It is in the Netherlands, that the assumption of the ideal-typical model describes the role of consensus in regulating animal biotechnology most clearly. During the legislative process, it was emphasized that no consensus existed yet on the extent and meaning of the intrinsic value of animals.[16] The lack of consensus was stressed during the parliamentary debate and in the Explanatory Statement of the regulation on animal biotechnology. The legislature stated that this concept needed further development, but should be included in the legal framework as an aspiration. The concept of intrinsic value of animals could thereby guide the process of deepening and strengthening the moral status of animals. A search for consensus was regarded as one of the mechanisms that would structure and actively stimulate this further norm development.[17]

8.2.5 Legitimacy

The interactive legislative approach emphasizes popular legitimacy as based on involvement of a broad spectrum of actors.

The interactive legislative approach emphasizes the substantive justification based on a critical reflection in light of principles, ideals, and policies.

Popular legitimacy was a means of validating legal norms in all three countries. However, the hypothesis related to ensuring the quality of law through substantive justification based on a critical reflection in light of principles, ideals, and policies could not be corroborated.

Section 5.3.3 identified two main sources of legitimacy for legal norms in the context of animal biotechnology in Denmark: legal validity and popular legitimacy. In addition, the regulation of animal biotechnology was justified substantively. Popular legitimacy is highly important in Denmark, as can be recognized in tools such as consensus conferences, citizens' panels and citizen juries. These instruments ensure public involvement, although indirectly, and contribute to establishing public support. Furthermore, a broad deliberative decision-making process in which all kinds of experts are involved helps to establish popular legitimacy. Thus, popular legitimacy does not only depend on public support, but also on broad deliberation.

Besides popular legitimacy, the Danish legal system seems to acknowledge the need for the substantive justification of legal norms. As extensively discussed in Section 5.3.3, Danish substantive justification is founded on a principle of proportionality. This principle in itself is content-less and largely procedural, as it prescribes that the content of decisions must be justified by a balancing of pros and cons. The AEC also adhered to this principle of proportionality in their reasoning on the moral impact of the animal biotechnology regulation. The Danish legislature uses this principle mainly for practical reasons, as it functions as a political tool for balancing pros and cons.

16 Section 7.2.1.
17 Section 7.2.2.

In Switzerland, legal validity plays an important role in constituting legitimacy and ensuring the quality of law. Official rules ensuring legal validity rely heavily on popular approval. Legal validity and popular legitimacy are therefore correlated in the Swiss system. In the legislative process, instruments such as referenda ensure public involvement. These referenda guarantee the popular approval of legal norms. In reality, referenda also help to ensure broad public involvement: the threat of a referendum is often enough to ensure that the opinions and concerns of the people are taken into account by legislators. Popular approval is prominent in other elements of Swiss political society as well. For example, the pre-parliamentary consultation procedure, in which the public is represented by various interest groups, is a tool for establishing broad involvement in decision-making.

In addition, consensus plays a prominent role in ensuring the legitimacy of legal norms. In some legal systems, particularly those that have elements of deliberative democratic traditions, consensus is related to the substantive justification of legal norms. However, in the Swiss system, consensus mainly ensures popular support.

Furthermore, the Swiss legislature has also acknowledged fundamental principles as a basis for legal norms. For example, the dignity of living beings played an important role in developing the legal norms on the regulation of animal biotechnology. Although such principles provide substantive justification for the law, the Swiss government must still rely on satisfying the people and ensuring their approval. It were the Swiss people who called for protecting the dignity of living beings. Here, we can recognize the expressive function of law as it reflects values important for the people.

In the Netherlands, both popular support and substantive justification are important for establishing the legitimacy of legal norms. With respect to the regulation of animal biotechnology, policy-makers were confronted with difficulties in establishing popular legitimacy. Due to the complexity of the problem, it was not possible to reach a substantial consensus. In response, Dutch policy makers established a process of broad deliberation with experts and the public and set up a system for disseminating information.

Substantive justification, though, was mainly found in the Dutch 'no, unless' policy concerning animal biotechnology. Like the Danish principle of proportionality, this principle is in itself content-less, but offers a procedural method for justifying the content of decisions. This policy reflects a restrictive approach, and should be understood in terms of a prohibition of biotechnological procedures with animals, unless important societal interests are at stake. The policy is based on extensive deliberation about the moral impact of animal biotechnology in which various values and ideals were discussed. The variety of values and the lack of consensus about their meaning and extent justifies a restrictive policy. Even though a substantial consensus on norms and values was lacking, this policy permitted a critical reflection on various moral considerations. Moral reflection thus had a central role in the legislative procedure as well as in the legal discourse that guides further norm development. The 'no, unless' policy was substantively justified by the various reports of experts on the moral considerations and values at stake, and the fact that it allows for moral reflection.

8.2.6 Long-Run Problem Solving

The interactive legislative approach focuses explicitly on long-run problem solving, understood in terms of coping adequately with the characteristic issues of a problem rather than reaching a fixed aim.

The regulatory strategies of the Danish, Swiss and Dutch legislatures only partly corroborated this hypothesis. However, in Switzerland and in the Netherlands this hypothesis was more explicitly corroborated than in Denmark. For animal biotechnology, 'long-run problem solving' requires a regulatory framework that can cope with rapid changes and moral controversies that have not yet crystallized. 'Solving', here, does not mean that a problem ends or a dispute is settled, but rather that a strategy has been developed that makes it possible to cope with the issue. Long-run problem-solving strategies are strongly related to room for norm development, since flexibility creates possibilities for addressing new issues and trends. In the three case-study countries, the legislators chose a licensing procedure for dealing with issues concerning animal biotechnology. This contributes to problem solving in the long run in terms of addressing the issue, since each new application involves a new reflection on the controversies. The case-by-case approach makes it possible to include new insights in this reflection. Therefore, licensing procedures seems to leave room for further norm development. However, the three countries do not all interpret these licensing procedures as a communicative framework for further norm development.

Especially in Denmark, the legislature seems to have closed off avenues for further development and for responding to new controversial trends. The licensing procedure concerning animal biotechnology adopted the licensing procedure for animal experiments.[18] It uses a standardized procedure that balances animal welfare criteria against the consideration of research aims. Licensing involves applying legal norms and standards, but not a process of norm development. This approach creates for the Danes a well-functioning procedure that can address current issues. However, it does not develop norms for future trends or deepen the understanding of more fundamental issues in the legal context. At the same time, adjusting the accepted research purposes for biotechnological procedures with animals does leave room for various technological developments. New technological developments that have similar purposes can be fit into this framework. Nevertheless, 'the long run' is limited to addressing new trends that are not fundamentally different. New technological applications with fundamentally different issues will require new regulation.

In Switzerland, it seems at first sight that the legislature developed a regulatory framework for licensing that aims to address problems in the long run. The use of open norms that explicitly require reflection on risks and moral controversies leaves room for further development. And the licensing procedure as such may contribute to coping with long-run issues concerning biotechnology. The ECNH has contributed to the development of a

18 Section 5.2.2.

set of standards for addressing the dignity of living beings and other new trends and controversial issues. And it can restrict itself to advising only about controversial issues and new trends, thereby avoiding the trap of being limited by the legal context. Nevertheless, in reality, development of the legal norms and especially of the moral issues is restricted.[19] The ECNH's standards were developed in a relatively early phase of the norm development process, and bypassed more fundamental questions about the meaning of concepts such as 'dignity of living beings'.[20] Thus, new trends are only addressed within the current framework of standards for licensing, which are not flexible to fundamental changes.

In the Netherlands, the legislature has attempted to provide a regulatory framework for coping with animal biotechnology in the long run. The legal framework consists of norms that are not just applied during the licensing procedure, but are also open for further development. The open norms with their basis in moral values express the fundamental issues at stake and at the same time guide further communication. The CAB aimed at further norm development, and has to a certain extent contributed to the development of new norms for addressing animal biotechnology in the long run.[21] It has developed various standards for licensing, and pronounced itself on the moral acceptability of some research. However, norm development remains restricted to the development of standards in the legal context, leaving the more fundamental moral questions unaddressed.[22] Consequently, further development has stagnated.

At the same time, the Dutch legislature interpreted long-run problem solving not in terms of developing a *strategy* to deal with ongoing controversies, but rather by seeking a *solution* to settle the controversial issues. In reality this interpretation involved that the legislature did not aim to stimulate ongoing norm development, but instead the regulation of animal biotechnology was considered only a temporary solution. As soon as the controversies were settled, the regulation could be placed out of order. With respect to the regulation of animal biotechnology, it sought this solution in insights into the (moral) acceptability of biotechnological procedures with animals, hoping that new understandings would settle the dispute and solve the problem. The Dutch legislature, here, acknowledged the temporary status of the regulation on animal biotechnology. Moments for re-evaluation were built into the regulatory framework. Indeed, the Decree on Animal Biotechnology is currently being discussed. The Dutch Minister has argued that the decrease in norm development should be interpreted as a closure of the debate. Closure indicates that people have reached an agreement about the moral acceptability of the current approach and that there have been sufficient insights into the controversies. And thus, further development is no longer necessary. The CAB's terms of reference have also changed since 1 July 2009, such that the CAB no longer plays a role in the licensing procedure. Medical applications of animal biotechnology, which make up most of the licensing applications, no longer require a license, as they seem morally accepted.

---

19 Section 6.2.2 and Section 6.3.2.

20 Section 6.3.2.

21 Section 7.2.2 and Section 7.3.2.

22 Section 7.3.2 and Section 8.2.2.

8.3 Conditions

One of the main research-questions addressed in this book is whether it is possible to identify the general conditions under which an interactive legislative approach could function well. Chapter 4 translated this overarching question into various sub-questions concerning the political cultures of the three case-study countries. These questions focused on whether the adequacy of the interactive legislative approach depends on the democratic traditions and legal culture of the country in which it is employed.

8.3.1 Democratic Traditions

This section discusses the democratic traditions of the three countries in light of the proposed conditions for an interactive process of decision-making. Democratic traditions mainly relate to the interactive element of the interactive legislative approach, since democratic traditions influence a country's conception of popular legitimacy and of the relation between government and public.

Denmark has the strongest orientation towards deliberative democratic traditions. In Denmark, decision-making is most obviously structured by a process of deliberation among various actors on a horizontal level. The Danish government uses various participatory methods to bridge the gap between scientists, government, and the people.[23] Although critical remarks can be made about the extent and effectiveness of these methods, people at least feel involved in decision-making. In the case of animal biotechnology regulation, it was primarily experts that were involved in decision-making. The public played only an indirect role, via representation and public debate. Nevertheless, expert involvement ensured that various actors were part of the deliberation. Experts were consulted for advice on drafting legislation. Furthermore, both experts and representatives of interest groups were appointed to the Animal Experiment Directorate to decide on licensing biotechnological procedures with animals. Despite the lack of direct public involvement in decision-making, therefore, deliberation still seemed to accommodate the conditions for an interactive process.

The Swiss political culture is a mixture of direct democracy and consociational democracy. Traditions of direct democracy have resulted in a set of various popular rights. However, these popular rights do not establish a horizontal level of communication among the actors involved. Direct democracy in itself is not sufficient to accommodate the conditions for an interactive process. Instead, it is due to the traditions of consociational democracy that broad participation on a horizontal level can take place. For example, the pre-parliamentary consultation procedure presents a norm development process characterized by broad participation in which various experts and interest groups can comment on regulation as well as suggest changes or amendments.[24] Indirectly, direct democracy contributed

23 Section 5.4.2.

24 Section 6.41. and Section 6.4.2.

to establishing the conditions for an interactive process, as it was the threat of a call for a referendum that brought about the shift towards a more consociational style. Taking the considerations, preferences, and interests of the various actors into account in an earlier stage of the legislative process could help to avoid a call for referendum. At the same time, however, direct democracies run the risk of being blind to the needs of new minorities. New minorities have neither the knowledge nor the capacity to call for a referendum or launch a popular initiative. Therefore they run the risk being overruled by majorities or traditional minorities that are better organized.

This blind spot, by which direct democracy excludes new minorities, affects the process of consociational decision-making as well. Consociational processes also risk excluding new minorities that are neither represented nor organized into groups that participate in decision-making. The extent of broad involvement based on co-operative effort on a horizontal level is consequently limited.

Dutch political culture is mainly characterized by elements of consociational democracy due to phenomena such as pillarization and the polder model. The latter reflects the co-operative effort of the interactive legislative approach, in which various actors participate in decision-making on a horizontal level. At the same time, the polder model is criticized for its back-room politics, which reveals a more hierarchical structure.[25]

The tenets of the polder model, which sets up a horizontal process of decision-making, can be recognized in the legislative process that led up to the regulation on animal biotechnology. This process was characterized by a broad process of horizontal participation. Furthermore, various actors can participate in the Dutch licensing procedure. The CAB, as an expert committee, has an advisory role in decision-making. Besides, the licensing procedure is open to public objections, which are presented in a public hearing chaired by the CAB. However, the CAB's role remains hierarchical in relation to the public, as in reality the CAB has a decisive role in licensing. The extent of broad involvement is thus restricted.

Despite restrictions on broad involvement due to political reality, the claims related to the interactive element of the ideal-typical model were in all countries to a certain extent corroborated. We can thus conclude that both consociational democracies and deliberative democracies can accommodate processes in which various actors can participate in decision-making. At the same time, we could question whether consociational democracies run the risk of excluding minorities that are not represented or organized into groups that participate in the consociational processes of decision-making. The Swiss case study is an example of this risk. Direct democracy by itself is less suitable for accommodating an interactive process, as it mainly provides the right to question decisions, not to deliberate about these decisions.

8.3.2 Legal Culture

The possibilities for a dynamic process of norm development in the legal context also seem to depend on a country's legal culture. As van der Burg and Brom argue, different

25 Section 7.4.1.

paradigms for addressing complex moral issues also influence conceptions of law. We should therefore ask: does a country's concept of law allow room for ongoing norm development after the implementation of legal norms? Can the 'methods' for establishing a dynamic process be incorporated into the country's concept of law?

In Denmark, the most influential legal theory – Scandinavian legal realism – finds its roots in legal positivism. Alf Ross's ideas, inspired by American legal realism, have resulted in an earthbound, pragmatic and realistic conception of law. Validity of law is measured in terms of effectiveness and probability of future application. Law is considered to be an instrument for establishing policy goals. Consequently, clear and concrete norms are preferable. Open norms, moral values, and alternative or additional functions of law cannot be justified within this positivist framework. Thus, a licensing procedure was the most pragmatic solution for a controversial issue like animal biotechnology. Licensing is a means for applying norms in the field, not for developing these norms any further. Denmark's positivist ideas about the concept of law explain the lack of dynamic interpretation in the Danish system. We can conclude that a view of law rooted in legal positivism and legal realism is not suitable for accommodating a legislative approach based on a dynamic process of norm development.

Swiss legal culture also has its roots in legal positivism. However, unlike in Denmark, the Swiss legislature deviated from the standard legislative approach and adapted a process-based approach for addressing the complex issues concerning animal biotechnology.[26] This process-based approach seems at first sight similar to the interactive legislative approach, and makes room within the Swiss conception of law for moral values. Nevertheless, the similarities between the Swiss process-based approach and the ideal-typical interactive legislative approach need to be put into perspective. First of all, there are some differences relating to the legislature's reasons for using a process-based approach. These reasons were mainly pragmatic and related to public support. Second, the legislature did not adapt a purely process-based approach. Instead, they used a mixture of the process-based and traditional product-based approaches. Errass stated that the legislature preferred a product-based approach, which indicates a positivist view of law. A process-based approach was easily assimilated into a product-based approach due to a preference for the latter. A preference for clear and concrete norms does not correspond with an aim for further development. Even though legal positivism is less dominant in Swiss legal culture than it is in Denmark, the influence of legal positivism still seems to undermine the objective of establishing a dynamic process. A positivistic view of law makes the interactive legislative approach a high-risk one that can easily fall back into focusing on clear and concrete norms.

Dutch legal culture is influenced by a mixture of legal theories.[27] Elements of interactionism are most prominent. Interactionist theories hold that legal norms concretize ideals and principles by applying the law in reality. Legal norms are developed through

26 Section 6.2.1 and Section 6.4.3.
27 Section 7.4.3.

interaction among the actors involved based on a shared interpretation of values. This seems to accommodate the elements of both interaction and dynamics. And indeed, for regulating animal biotechnology, the Dutch legislature aimed to establish a dynamic process of norm development. This is exemplified by the objectives of the CAB as well as by the use of open, aspirational norms.

The identification of these elements of an interactive legislative approach in the Dutch legislative practice seems to indicate that systems with roots in theories of interactionism can accommodate interactive legislative approaches. However, the Dutch regulation of animal biotechnology ultimately failed to establish an ongoing process of norm development. Norm development concerning the moral acceptability of animal biotechnology has not reached a higher aggregation level, but has instead been limited by the legal context.[28] One explanation can be found in the instrumental approach that is particularly preferred by the Dutch Minister of Agriculture. As in the Swiss case, preference for an instrumental approach turns the interactive legislative approach into a high-risk one that could easily fall back into traditional preferences for clear and concrete norms or standards. A second, more general explanation is that the conditions for a dynamic process must not depend solely on a country's legal traditions and its legal-theoretical roots. Interactionism seems to accommodate dynamic processes. In reality, however, despite the Netherlands' roots in interactionism, interactive dynamics were restricted.

The 'right' legal context is therefore not a sufficient condition for establishing dynamic processes. In legislative practice (particularly in positivist traditions), law and lawyers have a tendency to be rigid and to offer a clear solution to a problem as soon as possible. Law and lawyers have a tendency to structure practical thinking by a legalistic context. Due to this tendency, even the more dynamic legislative approaches risk falling back on offering a premature 'solution' of clear and static norms or standards. This risks simplifying the issue and, consequently, its solution. We could therefore question: to what extent can we open up the legal context to dynamic processes?

8.3.3 Ethics Committees

This section presents remarks relating to the role of ethics committees in addressing moral controversies. These remarks do not fit the framework of the ideal-typical model, but are nevertheless relevant for analyzing the interactive legislative approach.

Ethics committees figured prominently in the analysis of the three case studies. During the legislative process, the legislatures of all three countries consulted ethics committees for advice on the moral impact of animal biotechnology. These committees contributed to the development of norms for addressing the moral controversies associated with these issues. In the Netherlands and Switzerland, ethics committees also played an important role in licensing. The Dutch and Swiss committees were appointed to assess the moral impact of biotechnological procedures with animals and the moral status of animals. These

28 Section 7.4.3.

committees exemplify the interactive element of the norm development process. They could also play an important role in establishing a dynamic process.

At the same time, however, their role was questionable. The CAB faced difficulties in stimulating an ongoing process of norm development regarding the moral acceptability of animal biotechnology, particularly because its moral reflections were limited by the legislative context. The ECNH was appointed to promote public debate, but in reality restricted itself to providing information to the public. This left the public without a channel for making their voices heard.

8.4 Difficulties for Dynamics

Comparing the outcomes of the case studies with the assumptions and hypotheses of the ideal-typical model of the interactive legislative approach has resulted in some interesting outcomes. These outcomes challenge the adequacy of some of the assumptions of the ideal-typical model, especially those concerning the dynamic element of the interactive legislative approach.

To start, it seems that the interactive legislative approach was not operative in Denmark. The Danish legislature did not attempt to stimulate an ongoing process of norm development. Therefore, it lacks one of the basic tenets of the interactive legislative approach. In Denmark, the claims concerning the role of moral values and the use of open norms were not corroborated. Furthermore, the emphasis on consensus led to a closure of debate. Consensus was considered an aim of norm development, which ended with the constitution of the regulatory framework. Therefore, the outcomes of the Danish case study are less relevant for challenging the practical value of the interactive legislative approach, and less useful for testing its assumptions and hypotheses. Nevertheless, the Danish case study is still relevant for modelling the ideal-typical model since it contributed to identifying conditions under which an interactive legislative approach may be adequate. First of all, the interactive element of the approach could be recognized in the Danish approach to animal biotechnology as well as its general ideas about democracy and regulation. In general, deliberative traditions together with consociational democratic traditions seem to accommodate an interactive norm development process. Second, the Danish results provide insights into which jurisprudential traditions are less likely to accommodate an interactive legislative approach: legal positivism.

In Switzerland and the Netherlands, the claims of the ideal-typical model were only partly corroborated. The legislatures in both countries used open norms to stimulate an ongoing process of norm development. Furthermore, both aimed to ensure a role for moral values in all stages of norm development. The search for consensus, which was not yet established when constituting the regulatory framework, was considered a purpose of norm development. And both national legislatures presented regulation as a long-term approach for problem solving. The aims of the legislatures as well as these characteristic elements of the regulation of animal biotechnology correspond with the descriptive assumptions and hypotheses of the ideal-typical model.

However, despite the fact that most of the ideal-typical model's claims were corroborated, an ongoing process of norm development was not established in either country. Instead of becoming mutually-reinforcing, both legal and moral norm development processes have stagnated. Other difficulties relate to the role of consensus and long-run problem solving. In Switzerland, consensus was not only considered an end, but also a starting point. Besides, in all three case-study countries, aiming for consensus did not contribute to establishing an ongoing process. Furthermore, in the Netherlands, long-run problem solving was interpreted in terms of settling the dispute. The long run was therefore limited to a certain time period. In reality, moral values did not play a role in all stages of norm development, since norm development stagnated prematurely. The use of open norms did not stimulate an ongoing process.

These remarks challenge the adequacy of the hypotheses describing the ideal-typical legislative approach, as well as the claims about the possibility of stimulating an ongoing process of norm development. However, the model was not weak throughout. The results of the case studies do not call into question whether interaction and dynamics are an adequate method for dealing with complex issues with a strong moral impact. The focus on interaction seemed to correspond with the approaches in the case studies. Broad involvement based on a horizontal co-operative effort resulted in a regulatory framework that was close to reality. Furthermore, it established satisfaction among most actors since it created an open forum in which considerations were discussed and challenged.

The hypotheses and assumptions that were not corroborated or did not have the assumed effect mainly involve the dynamic aspects of norm development. Analysis of the jurisprudential traditions of the case-study countries shows that the conditions for dynamic approaches depend not only on a country's legal roots, but also on other factors. The Dutch legislature aimed for a dynamic process, and the Netherlands' roots in interactionism would seem to accommodate all elements of the interactive legislative approach. However, a dynamic process was not established. How should we understand this? The dynamics of the approach relate to the difficulties of reinforcing moral and legal norm development and the assumption about consensus. The outcomes of the case studies therefore challenge the appropriateness of these two methods of establishing an ongoing process. In Chapter 9, I will focus on a critical analysis of the role of consensus, as it seems that aiming for consensus is one of the features that challenge the dynamics of the interactive legislative approach. My case-study outcomes reveal a blind spot for counteracting debate. In Chapter 10, I will additionally focus on the shortcomings caused by a dominant legalistic context in which the decision-makers have to operate. The legalistic context seems to involve a second challenge to the dynamics of the interactive legislative approach. These challenges will be addressed in Part III.

Part III

9 Consensus Reconsidered

9.1 Introduction

This chapter reconsiders the role of consensus for stimulating an ongoing process of norm development in the interactive legislative approach.

Consensus plays an important role in deliberative democracy theories. In general, two conceptions of consensus can be distinguished. The first presents consensus as an ideal outcome. Consensus in this sense is a static concept. The idea of consensus as an ideal outcome is derived from Habermas' idea of the *ideal speech situation*. Habermas presents a normative theory of a procedural democracy in which the ideal political process is one that aims for consensus under ideal circumstances. Under ideal circumstances, all discussants can participate in debate with equal opportunity and equal power, and without any constraint. If all parties consent to an outcome at the end of this discussion, it should represent the ideal outcome.[1] Thus, consensus is the ideal outcome of the ideal political process.

The second conception understands consensus as a regulative ideal. Consensus in this sense is a dynamic concept. Here, consensus is not the ideal outcome of a political process, but rather an orienting aim. Proponents of consensus as a regulative ideal consider the outcome as less foundational for their decisions. They acknowledge that in legislative politics a consensus may never be reached. Instead, they emphasize the benefits of aiming for consensus as a method for structuring deliberation. Gutmann and Thompson, for example, consider aiming for consensus as an ideal outcome, but consider aiming for consensus only as a method of reducing disagreements and ensuring openness to other people's viewpoints.[2] Consensus, then, functions merely as a regulative ideal for rational debate in which discussants persuade using rational arguments.

The ongoing process of norm development outlined in the ideal-typical model of the interactive approach must be understood in the light of consensus as a regulative ideal. As is explained in Chapter 3, Van der Burg and Brom's original model does not presuppose consensus. Instead, it argues for a constructive interaction aimed at consensus. By the same token, aiming for consensus will stimulate this constructive interaction in terms of promoting an ongoing process of norm development.

---

1 J. Habermas, *Structural Transformation of the Public Sphere: An Inquiry into a Category of Bourgeois Society*, translated by T. Burger and F. Lawrence, Cambridge, MA, MIT Press, 1989; See N. Rescher, *Pluralism, Against the Demand for Consensus*, Oxford, Clarendon Press, 1993, p. 190.

2 A. Gutmann and D. Thompson, *Democracy and Disagreement*, Cambridge, MA, Harvard University Press, 1996.

As the case studies showed, however, aiming for consensus in the regulation of animal biotechnology has only partly established an ongoing process of norm development. The mutually reinforcing development of moral norms and legal norms was restricted to the initial stages of norm development. In all three countries, the development of both types of norms stagnated prematurely. This raises questions about consensus as a regulative ideal: can the search for consensus adequately stimulate an ongoing process of norm development? This chapter will address this question by reconsidering consensus as a regulative ideal. In doing so, it will also elaborate the role that consensus played in the case studies, which was not always similar to the role of consensus envisioned in the ideal-typical model.

Section 9.2 consists of two sub-sections. First, it will explain the context in which this reconsideration of consensus must be placed. This reconsideration is restricted to a critique of the adequacy of consensus for addressing complex moral issues within an interactive legislative approach. This context also shows that my critique is focused on a specific definition of consensus. Therefore, second, this section will present a definition of consensus that is similar to consensus as a regulative ideal. In defining consensus this section will also briefly various understandings of consensus, in particular consensus in the Habermasian definition of consensus as an ideal outcome. Exploring the Habermasian definition of consensus will contribute to come to a better understanding of the critiques against consensus as a regulative ideal.

Section 9.3 presents a critical reflection on the use of consensus in interactive legislative processes. This section will describe an 'ethos of consensus': a way of thinking in legal politics that seeks to structure norm development and decision-making processes around a search for consensus. The ethos of consensus includes both consensus as a regulative ideal and consensus as an ideal outcome. To begin with, this section criticizes the ideal-typical models' assumptions about consensus. Then, it discusses the outcomes of the case studies, which show the shortcomings of consensus as a regulative ideal in legal practice.

Finally, Section 9.4 presents a reconsideration of consensus as a regulative ideal and sketches some preliminary ideas for an alternative. This alternative will be further explored in Chapter 10.

9.2 Preliminary Remarks

9.2.1 *The Context of Reconsideration*

Before beginning the critique of consensus that makes up the bulk of this chapter, it is necessary to first discuss the context in which this critical reflection must be understood.

To begin with, the critique presented here is restricted to the role of consensus in the interactive legislative approach. The model interactive legislative approach is strongly focused on consensus; in the terminology adopted here, this legislative approach follows an ethos of consensus. The ideal-typical model of the interactive legislative approach aims for consensus in two senses. First, it sees consensus as a method of stimulating ongoing norm development processes. And second, it uses consensus a justification for these processes. In this approach, the development of legal norms and development of norms in practice are supposed to be mutually reinforcing. The search for consensus plays a role in the interaction between these processes of norm development. However, it can be questioned whether, instead, a strong focus on consensus is counterproductive.

Second, the criticism is restricted to the use of consensus as a guiding norm for particular kinds of issues. The interactive legislative approach aims to cope with complex policy issues that give rise to strong moral concerns. Animal biotechnology issues are characterized by intractable moral disagreements. The difficult nature of these issues complicates reaching consensus. It should be questioned whether consensus is possible for this type of issues, and, thus, whether they should adopt consensus as an action-guiding ideal at all.

9.2.2 *Defining Consensus*

This section starts by defining consensus both as a regulative idealand as the ideal outcome. In the ideal-typical model of the interactive legislative approach consensus is not assumed to represent the ideal outcome. Nevertheless, discussing the conception of consensus as the ideal outcome helps to flesh out the critiques against consensus presented in the next section. Consensus as regulative ideal and consensus as ideal outcome are related in two ways. First of all, consensus as a regulative ideal is intertwined with consensus as the ideal outcome, and therefore falls prey to some of the same critiques. And second, the case studies have shown that consensus as a regulative ideal runs the risk of being counterproductive because it leads to decisions being presented as though they were based on consensus. Temporary definitions and solutions defined at the outset of the norm development process are not further challenged, and in reality are applied *as if* based on a consensus. In other words, consensus as a regulative ideal leads to the use of fictions of consensus. The fictional use of consensus results in the stagnation of norm development. Furthermore, fictions of consensus are problematic because they function as a false basis for decision-making. In the case studies, consensus functioned not only as a regulative ideal, but also as a source of legitimacy.

Consensus as a Regulative Ideal

In the ideal-typical model of the interactive legislative approach, 'consensus' means a quest for consensus. According to Van der Burg and Brom, it would be enough if the various actors involved could freely deliberate together to come up with norms that are orientated to practice. If norm development processes are structured as quests for consensus, they will be more likely to establish a dialogue with the practice or the field. For Van der Burg and Brom, consensus is not a starting point, but rather one of the purposes of debate. They point out that this will 'hopefully lead to a consensus or at least to clarification of the differences in opinion, which may be the starting point for a compromise'.[3]

3 Van der Burg and Brom, 'Legislation on Ethical Issues', p. 64.

Similar conceptions of consensus as a regulative ideal can be found in deliberative theories.[4] In deliberative theories, aiming for consensus is an incentive for establishing rational debate. Here, the value of aiming for consensus does not (only) depend on achieving consensus as a final outcome.[5] Gutmann and Thompson, for example, value the open attitude that aiming for consensus brings about. Furthermore, the commonality of the search for consensus helps to establish the discourse in which the debate takes place.[6] Without a common aim a discourse can be difficult to begin. According to Gutmann and Thompson, without commonality in political discourse it is nearly impossible to constitute a forum in which all actors have an equal say and all can actually deliberate together.

The definition of consensus that is used in the interactive legislative model is consensus as a regulative ideal. Van der Burg and Brom's ideas about the role of consensus show sufficient similarities with the ideas of deliberative theorists to justify this definition. Defining consensus in this way allows for a better elaboration of the elements of the critique in the subsequent sections.

Consensus as the Ideal Outcome

Another way of thinking about consensus is as the ideal outcome of a legislative process. This conception is based on Habermas' understanding of consensus. Habermas explains consensus as the ideal outcome of deliberation under ideal circumstances. Habermas argues for a discourse theory with normative connotations of democracy. Discourse theory emphasizes the institutionalization of procedures and conditions of communication that could form the basis for the success of deliberative politics.[7] Habermas argues for a procedural democracy with its normative ideal in the ideal speech situation. Habermas acknowledges that consensus is what he calls a 'counterfactual ideal', which may never be reached. Cohen goes a step further, and acknowledges that in political reality a consensus may never be reached at all. Nevertheless, Cohen argues that the search for consensus can encourage participants to be more open to other people's viewpoints. This openness to other viewpoints causes participants to also be open for persuasion by the arguments of others. Deliberation under these circumstances will lead to decisions that are more morally legitimate ideal outcomes.[8]

Here we see that consensus as an ideal outcome is interwoven with the benefits of aiming for consensus. Consensus as an ideal outcome cannot be completely separated

4 See for example Gutmann and Thompson, *Democracy and Disagreement*; and J.D. Fearon, 'Deliberation as Discussion', in J. Elster (ed.), *Deliberative Democracy*, Cambridge, Cambridge University Press, 1998, pp. 44-68.

5 Cohen, 'Deliberation and Democracy', p. 75.

6 Van der Burg and Brom, 'Legislation on Ethical Issues', pp. 57-75.

7 J. Habermas, 'Three Normative Models of Democracy', in S. Benhabib (ed.), *Democracy and Difference. Contesting the Boundaries of the Political*, Princeton, Princeton University Press, 1996, pp. 21-30.

8 J. Cohen, 'Deliberation and Democratic Legitimacy', in J. Bohman and W. Regh (eds.), *Deliberative Democracy. Essays on Reason and Politics*, Cambridge Mass, The MIT Press, 1997, pp. 67-91; see also C. Gould, 'Diversity and Democracy: Representing Differences', in S. Benhabib (ed.), *Democracy and Difference. Contesting the Boundaries of the Political*, Princeton, Princeton University Press, 1996, pp. 171-186.

from the process of aiming for consensus. Aiming for consensus, although it may never be reached at all, is most likely to create circumstances that can legitimate decisions made during this process. Legitimacy thus depends on the process of deliberation that is followed.[9] Consensus as the ideal outcome still forms the main focus, however. Creating the ideal circumstances (by aiming for consensus) is good because it will contribute to reaching the ideal outcome (a decision based on consensus).

The understanding of consensus as an ideal outcome is often criticized. For example, scholars like Rescher contrast it with the value of pluralism in the political context.[10] Rescher's argument builds on a connection between the desirability of and the possibilities for reaching consensus as the ideal outcome. He argues that consensual positions view the social order as a utopia. Rescher criticizes Habermas's ideal speech situation in which consensus is a touchstone of political legitimacy. He outlines two shortcomings of this approach. The first shortcoming concerns the idealization of the method, which is far from reality. The second one relates to the gap between idealized consensus and real-world problems. According to Rescher, this gap is too large to bridge. In his book *Pluralism* he argues that we need to acknowledge that we live in an imperfect world. Our (intellectual) resources are limited. And beyond that, we have to be aware of pluralism in matters of opinion. It is essential to learn to live with and acknowledge dissensus. The toleration of dissensus will prove to be constructive and positive in settings of social interaction.[11]

To explain his argument, Rescher distinguishes between ideals and idealization.[12] Ideals are action-guiding: they define something valuable that we should strive for. They are desirable goals that are evaluated positively. Idealizations, on the other hand, are of no practical use. Instead, they are theoretical concepts that can be thought about, but in reality cannot be reached. They are 'thought-instruments' that may be theoretically profitable to consider, but are senseless to actually aim for in reality.[13] While idealizations may be positive and desirable, therefore, they are not plausible action-guides. According to Rescher, unrealizable idealizations cannot function as action-guiding ideals: '[..] a preoccupation with the unrealizable ideal can do real damage in this domain where a pursuit of the unrealizable best can all too easily get in the way of the realization of attainable betterments and impede the achievement of realizable positive objects'.[14] He argues that consensus is an idealization, since it 'prescinds from the variety, diversity, and dissonance that inevitably characterize the beliefs, opinions, goals, and values of any sizable human community'.[15] According to Rescher, consensus is, thus, far from reality. Searching for a consensus that is unrealizable stands in the way of alternatives. According to Rescher, therefore, consensus is not a valid ideal.

9 S. Benhabib, 'Toward a Deliberative Model of Democratic Legitimacy', in S. Benhabib (ed), *Democracy and Difference. Contesting the Boundaries of the Political*, Princeton, Princeton University Press, 1996, pp. 67-94.
10 Rescher, *Pluralism*, Ch. 10.
11 *Ibid.*, pp. 1-4.
12 *Ibid.*, Ch. 10.
13 Rescher, *Pluralism*, p. 196.
14 *Ibid.*, p. 198.
15 *Ibid.*, p. 196.

Rescher's ideas represent an interesting and convincing analysis of the difficulties of positioning consensus as the ideal outcome.[16] His critique seems relevant for issues like animal biotechnology that are characterized by a profound moral pluralism. However, Rescher's critique seems less relevant in the context of consensus as a regulative ideal, as this alternative conception does not focus on consensus as the ideal outcome. Without this emphasis on consensus as the ideal outcome, this interpretation seems to avoid the idealization of consensus. Understanding consensus as a regulative ideal therefore seems to address Rescher's critique. At the same time, however, we could question whether consensus as a regulative ideal also involves an idealization of the possibilities for open attitudes, and thereby falls prey to the same criticism.

Consensus as a Substantive Justification

The concept of consensus as an ideal outcome also appeared in the case studies as a source of substantive justification. Chapter 2 presented a framework of dimensions on which the quality of law may depend. In the ideal-typical model of the interactive legislative approach, popular legitimacy and substantive justification are the two dimensions that ensure the quality of law.

To begin with, consensus (as either the 'ideal outcome' or 'regulative ideal') can be a source of popular legitimacy. Popular legitimacy is related to the consent of the parties affected by a rule. Legitimacy is based on the consent of the people: it reflects *public* morality. In light of popular legitimacy, consensus also has an expressive function, as it indicates the people's viewpoints.

However, consensus conceived of as an ideal outcome can additionally act as a source of substantive justification.[17] Chapter 2's framework for the quality of law described how reflection on critical morality can provide a substantive justification of law. The idea of consensus as anideal outcome includes various positions and viewpoints that have been contested with regard to their critical morality. Consequently, a consensus outcome can justify these decisionssubstantively.

Chapter 1 presented animal biotechnology as an intractable moral disagreement.Consensus on this type of issue is difficult to reach, since diametrically opposed viewpoints cannot be included together in a single outcome. Consequently, it is difficult, if not impossible, to establish substantive justification for issues like animal biotechnology on the basis of consensus.

At first sight, this may seem unproblematic, as the model interactive legislative process does not assume that consensus will establish substantive justification or popular legitimacy. However, the case-study outcomes reveal otherwise. In each of the three countries, consensus played a role in justifying legal norms. Although legislatures acknowledged

16 However, his ideas must be seen in the right perspective as for some issues consensus can be reached between certain groups of people. Issues that may be less intractable may have common ground on which consensus can be built.

17 Habermas, 'Three Normative Models of Democracy', p. 27. Habermas argues in terms of morally legitimated.

that a consensus did not yet exist, further debate was built on a fiction of consensus. Temporary definitions and decisions were applied *as if* they were based on consensus. These fictional consensuses were problematic sources of substantive justification, and ran the risk of counteracting the process of norm development. Building on a fictional consensus has the tendency to close off avenues of further reasoning and exclude viewpoints that contradict this fictional consensus. People take these fictions for granted and do not question them further. In words of Brand-Ballard: 'The longer we wait before addressing possible inconsistencies in common morality, the more those inconsistencies will infect our developing solutions to hard cases. Those infected solutions, in turn, gain popularity. People initialize them.'[18] I will exemplify this by the case-study outcomes.

The form of consensus discussed in the case studies can be seen as a fiction of the 'ideal outcome'. At first, the Dutch and Swiss concepts of 'intrinsic value of animals' and 'dignity of living beings' were not presented as being based on either a consensus or as fictions of consensus. Agreement existed only about the fact that the issue at stake caused moral controversies. This was not even an 'agreement to disagree', but an 'agreement of disagreeing'. In both countries, it was clear that there were still many contested elements and that, consequently, a consensus on these concepts did not exist. In Switzerland, even the terminology itself led to controversies.[19] Despite the various controversies, legislators felt that these concepts could function as points of departure for further norm development and critical moral reflection.

At the same time, analysing the legislation in practice has shown that in both the Dutch and Swiss licensing procedures the minimal definitions were adopted and not further challenged.[20] In reality, therefore, they did not function as points of departure for further development, but instead functioned as ends of the norm development process. Here we can recognize a fiction of consensus: the minimum and temporary definition of the norms was applied *as if* it were based on a consensus. Because they were presented as representing a consensus, these minimum definitions and temporary outcomes were not further challenged. After all, if a consensus has been reached there is less need for further deliberation.

The Danish case study showed similar outcomes relating to the concept of 'integrity of animals'. During the legislative process, integrity was extensively discussed and interpreted.

Respect for the integrity of animals was included in the Danish legislation through a restriction of the purposes for which animal biotechnology could be used. Integrity of animals was not further explained, and there was no room for challenging the rules in light of new developments. In the balancing of pros and cons that resulted in the list of four acceptable purposes for which animal biotechnology could be used, the concept

18 J. Brand-Ballard, 'Consistency, Common Morality, and Reflective Equilibrium', in *Kennedy Institute of Ethics Journal*, 13/3, 2003, p. 255.

19 See Section 6.3.2.

20 See Section 6.3.2 and Section 7.3.2.

of integrity was applied as if it were based on a consensus. Consequently, the notion of integrity itself was not further challenged. The notion of integrity was not, however, actually built on a consensus. It did not include a critical reflection on the various arguments regarding the morality of biotechnological procedures. Furthermore, a clear and comprehensive definition of integrity of animals had not yet solidified.[21] At the same time, the Danish Minister felt that further definition of the concept of integrity of animals was not necessary.[22] Adopting an instrumental and pragmatic approach, the Minister aimed for clear and concrete norms. A balancing of pros and cons to come to applicable standards was, for his purposes, sufficient.

Here, we see the implications of positioning consensus as an ideal outcome. Consensus implies that parties have worked out their differences. Therefore, it suggests that it is unnecessary to challenge this outcome over and over again.

However, it is not only consensus as an 'ideal outcome' that is problematic in this sense. The use of a fiction of consensus also complicates the use of consensus as a 'regulative ideal' in legislative practice. The use of fictional consensus limits the further development of norms to developing practical standards applicable in licensing. To this end, avenues for more general development were closed off prematurely. Furthermore, treating these concepts as if they were based on a consensus is mistaken.

Reconsidering consensus in this chapter therefore involves reconsidering the ethos of consensus as a whole. With respect to the interactive legislative approach, this ethos of consensus includes consensus as a regulative ideal and consensus as a source for substantive justification, particularly when it appears as fictional consensus. Nevertheless, reconsidering the ethos of consensus cannot be completely separated from consensus as the ideal outcome. Indeed, as we have seen in this section, the concepts of consensus as an ideal outcome and as a regulative ideal frequently collapse onto one another as a result of the tendency to present fictional consensus and use consensus as a source of substantive justification.

9.3 An Ethos of Consensus

An ethos of consensus is a way of thinking in legal practice that seeks to structure decision-making processes around a search for consensus. It includes consensus as an ideal outcome, as a regulative ideal, and as a fictional concept. Despite the differences between these concepts, their tendency to collapse onto one another means that they are best considered together as part of the same 'ethos of consensus'.

An ethos of consensus was present in all three case study countries, and had implications for the interactive legislative process in each of them. In Switzerland and the Netherlands, norm development was restricted to developing legal standards that were

21 See Section 5.3.2.

22 Therefore, the outcomes of the Danish case study will not be further elaborated in light of stagnation of norm development. After all, the Danish legislator did not aim for dynamics in the first place.

easily applicable in licensing practice. The norms themselves were not further concretized or developed to a higher aggregation level. One explanation is given in the previous section. The minimum definitions were adopted and not challenged further. These definitions were perceived as if based on a consensus. Their content was consequently supposed to be justified and not challenged further.

In Dutch and Swiss legislative practice, the presence of an ethos of consensus led to the premature stagnation of norm development. Temporary political achievements were not challenged further, even when it was clear that they required further development. The ethos of consensus undermined the possibilities for dynamic norm development and did not act as a stimulation for debate.

A second explanation can be found in the limits of aiming for consensus. Aiming for consensus may, to a certain extent, ensure positive attitudes towards decision-making processes and their outcomes. The common focus on reaching consensus in a particular case will often lead actors in the same direction. At the same time, however, the search for consensus seems to idealize this willingness to co-operate, and assume that it will consequently lead to rational debate or a further crystallization of norms. Willingness to co-operate does not imply that all parties consent on how to reach a common end. During the norm development process, differences in problem-definition, on the direction in which a discussion should move, as well as on new controversies can come to bear. Furthermore, aiming for consensus runs the risk of excluding controversies or differences. Actors that do not accept the problem-definition or the direction of the discussion may risk exclusion from the process.

The Swiss primate case exemplifies this risk. The ECNH interpreted dignity of living beings as a gradual concept. Licensing standards built on this conception. However, discussions concerning particularly controversial license applications involving biotechnological procedures with primates revealed that not all actors had consented to the gradual concept. During the legislative process, the lack of consensus on the definition of 'dignity of living beings' did not seem problematic, as the temporary status of the definition was emphasized. However, in licensing practice dissenting concerns and viewpoints were not re-introduced. The fiction of consensus on the gradual concept masked the fact that parties were still diametrically opposed, which resulted in heated debates.

It is, thus, doubtful whether positive attitudes will still exist at the end of debate. The excluded are no longer involved in development of the norms. Aiming for consensus that the exclusion of groups that may jeopardize the possibility of reaching consensus. Open and positive attitudes towards decision-making will be more difficult to sustain if dissenters feel that they are not taken seriously or if they have no space to challenge decisions. According to Honig, these groups will eventually withdraw from decision-making and radicalize.[23] Furthermore, exclusion undermines the rationality of debate, which in consensus theory is built on equal power and equal opportunity.[24]

23 B. Honig, *Political Theory and the Displacement of Politics*, Ithaca, NY, Cornell University Press, 1993.

24 See Rescher, *Pluralism*, p. 190.

An example of this can be found in the Dutch licensing procedure. In public hearings about licensing, questions about the moral acceptability of animal biotechnology were put aside with a simple 'this is not the forum for asking these questions'. The CAB sidelined arguments that went beyond their context of interpretation.[25] The public did not have equal opportunity or equal power to express their concerns, as the CAB did not respond to their objections using rational arguments.

Further inspiration for criticisms against consensus as an action-guiding ideal can be found in agonist political theories. Agonists doubt that consensus as such can be achieved through rational debate. In general, agonists criticize rationality and the institutionalization of politics and deliberation, as this does not correspond with political reality. The rationalization and institutionalization of politics results in the exclusion of grievances and concerns that may seem irrational. According to Mouffe, politics provides criteria for distinguishing the reasonable from the unreasonable.[26] In line with these doubts, agonists criticize consensus as an action-guiding ideal. They argue that politics is about contestation and conflict. Contestation challenges power relations, whereas aiming for consensus pacifies conflicts. Aiming for consensus pushes discussants to reach decisions and close the debate. Agonists argue that this closure takes away the possibility for people to understand their own viewpoints.[27]

Focusing on consensus tends to downplay the complexity of difficult issues. According to Rawls, the exclusion of viewpoints is the price we pay for addressing fundamental political questions. He argues that although it may be unfair to restrain ourselves and others from using non-neutral arguments, it is required for finding a solution that is acceptable to everyone in a pluralistic society.[28] However, the necessity of finding a solution cannot justify the exclusion of viewpoints. Rather, as Van der Sluijs et al. argue in the context of environmental sustainability, there should be more space in debates for diversity and uncertainty in knowledge and views. Van der Sluijs et al. suggest that making room for dissent will help provide a better picture of the status of climate science. Without a good picture it is difficult to develop prudent policies.[29]

As Wolthuis explains the Dworkin's right-answer thesis in the Hart-Dworkin debate, this thesis posited that in the context of judicial decision-making, a judge can only be convinced of the right answer if all interpretations have been contested. The exclusion of certain viewpoints without consideration therefore cannot be justified.[30] In the same way,

---

25 See Section 7.2.2, Section 7.3.1 and Section 7.3.2.

26 A. Schaap, 'Political Theory and the Agony of Politics', in *Political Studies Review*, 5, 2007, pp. 56-74.

27 Schaap, 'Political Theory and the Agony of Politics'.

28 See Grotefeld, 'Self-Restraint and the Principle of Consent', p. 87. He refers to J. Rawls, *Political Liberalism*, New York, Columbia University Press, 1996, pp. 29-35.

29 J.P van der Sluijs, R. van Elst, and M. Riphagen, 'Beyond Consensus: Reflection from a Democratic Perspective on the Interaction between Climate politics and Science', in *Elsevier Editorial System for Current Opinion in Environmental Sustainability*, 2010/2, p. 415 (409-415).

30 B.J. Wolthuis, 'Rechtsvorming als debat: De overtuiging het juiste antwoord te geven als voorwaarde voor goede rechtsvorming', in E.J. Broers and B. van Klink (eds.), *De rechter als rechtsvormer*, Den Haag, Boom Juridische Uitgevers, 2001, pp. 108-112.

we cannot be convinced that a consensus reflects the right answer if all possible viewpoints have not been incorporated. Besides, the right answer is more valuable if its foundations have been explicated and shared with others. Actors who may still doubt whether an outcome is the right answer for them may then be more willing to accept it Aiming for consensus has a tendency to excluding viewpoints. An ethos of consensus is therefore not a desirable feature in an interactive legislative process.

9.4 An Ethos of Consensus Reconsidered

9.4.1 *An Alternative?*

This chapter has criticized the search for consensus as a method of establishing a process of norm development. It has shown that following an ethos of consensus instead runs the risk of counteracting norm development processes. These criticisms explain the difficulties encountered in the case studies for successfully instituting a dynamic process. Because the regulations examined in each of the case-study countries all embodied an ethos of consensus, they were unable to constitute a truly interactive process of norm development. Coglianese arrives at a similar conclusion after reviewing empirical studies of negotiated rulemaking in the United States. She argues for alternative methods of participation in which agreement is not the main focus. According to her, striving for consensus is ultimately ineffective. Instead, a more realistic strategy would leave room for controversial viewpoints.[31]

The question, therefore, is: is there an alternative to an ethos of consensus? After all, decisions have to be made, and open norms have to be further developed. Is there an alternative for what consensus was presumed to establish? Is there an alternative that can allow open norms to function adequately and stimulate a dynamic process?

In the context of public deliberation, Bovenkerk suggests an alternative goal of revealing differences and pluralism.[32] When dealing with intractable moral disagreements it is essential to discuss and challenge all viewpoints openly. This will give people the opportunity to think through the essence of their viewpoints as well as to change them. Bovenkerk argues that revealing differences and pluralism should broaden the debate. This broadening of the debate should take place in a space called 'the agon'.[33] The 'agon' is a space in the public arena where conflict can exist and viewpoints can be challenged; where unreasonable and non-neutral arguments as well as reasonable and neutral ones can be expressed and debated. In the agon, the aims are broadening the debate, revealing differences, and exploring pluralism.

31 C. Coglianese, 'Does Consensus Work? A Pragmatic Approach to Public Participation in the Regulatory Process', in Alfonso Morales (ed.), *Renascent Pragmatism. Studies in Law and Social Science*, Aldershot, Ashgate, 2003, p. 182.

32 B. Bovenkerk, *The Biotechnology Debate*, p. 142.

33 *Ibid.*, p. 143.

Bovenkerk argues that decision-making processes should take place along a different track and with a different goal in mind.[34] Decisions have to be made, and a goal of revealing differences and pluralism will, then, be insufficient. The ground for decisions cannot be made based on differences and conflict. While Bovenkerk does not argue explicitly for consensus in decision-making processes, she still seems to hold on to the traditional ethos of consensus by advocating the use of 'incompletely theorized agreements'[35] or majority decisions as an alternative basis for decision-making when consensus cannot be reached. Even though consensus is not a requirement, therefore, it is still the preferred means of decision-making.[36] For example, incompletely theorized agreements are explained as a consensus about a practical decision: people may disagree on fundamental theoretical questions while achieving agreement or consensus about practical decisions.[37]

Bovenkerk's ideas for an institutional track of decision-making, however, prematurely abandon the goals of revealing differences and pluralism. She does not ignore the existence of controversies, but she leaves some loose ends about the role they should play. The institutional track she presents does not adequately reflect the fact that controversies may still exist after decision-making.

The next chapter explores an alternative: an ethos of controversies. An ethos of controversies presents an alternative for establishing dynamics in the process of norm development. It allows a role for conflict and differences in the decision-making process, without denying that decisions have to be made. At the same time, it is aware of the limits of dynamics in the legal discourse. Therefore, Chapter 10 recommends a two-track approach for stimulating ongoing norm development. It argues that an ethos of controversies can only address the failures of an ethos of consensus if it operates as part of a two-track approach.

---

34 *Ibid.*, pp. 144-145.

35 C.R. Sunstein, *Legal Reasoning and Political Conflict,* New York, Oxford University Press, 1996, pp. 35-61. Incompletely theorized agreements involve that an agreement is found on particular decisions, but without deciding on decisions about general theory or political philosophy. The basis of a decision is not theorized. These ideas follow from analogical thinking of agreeing on a low-level principle, while people do not agree on a general theory why this low-level principle may be sound. It is not so much about agreeing to disagree on the fundamental backgrounds, but it seems more a matter of agreeing to agree on some points and disagreeing on others.

36 Bovenkerk, *The Biotechnology Debate,* pp. 143-145.

37 Sunstein, *Legal Reasoning and Political Conflict,* p. 47.

10 An Ethos of Controversies in a Two-Track Approach

10.1 Introduction

This chapter presents an alternative to an ethos of consensus: an ethos of controversies. Chapter 9 explained and criticized the ethos of consensus that characterized the interactive legislative approach in the three case-study countries. It argued that aiming for consensus runs the risk of prematurely ending norm development, as temporary political achievements are treated as static ends. In the case studies, temporary achievements were applied as if based on a consensus. Aiming for consensus while at the same time assuming that a consensus has been reached will not stimulate further norm development. Chapter 9 suggested that to avoid those risks, an alternative to an ethos of consensus is required.

This chapter argues that an ethos of controversies can provide that alternative. An ethos of controversies can best be explained as a different way of thinking in approaching legal reasoning and decision-making. Instead of focusing on consensus, this ethos focuses on the controversies that characterize an issue. This focus on disagreement, also present in the theories of Waldron and the agonists, aids the process of norm development and assures that contrasting opinions and values are represented before, during, and after the decision-making process. This ethos of controversies is elaborated in three stages. First of all, controversies are articulated. Second, there is a confrontation between the various viewpoints in order to promote further norm development. Third, disagreements that may still exist even after decisions have been made are acknowledged.

Aiming for consensus is not the only difficulty for maintaining the dynamics in legal practice, however. Chapter 8 also pinpointed difficulties with respect to stimulating mutually reinforcing processes of moral and legal norm development. Reinforcement, here, implies a strong relation between legal and moral norm development in terms of combining and even intertwining these processes. This mutual reinforcement was intended to result in norms that would remain oriented to practice.

Instead, the case studies showed that moral norm development is often dominated by the legalistic context in which it takes place.[1] This chapter will argue that an ethos of controversies may address some of the failures of dynamics, but that it will still face difficulties due to the limits of legal discourse. As a result, this chapter suggests that it is necessary to acknowledge these limits in legal discourse. The ideal-typical model of the interactive

1 A legalistic context differs from a legal context. Both terminologies will be used in this chapter. A legalistic context refers to a context that is structured and framed by legal norms, legal principles, and legal concepts. The framework of discussion is then mainly a legalistic one. A legal context is the area in which legal decisions are made. Examples of legal contexts are the licensing procedures and legislative processes.

legislative approach was flawed, as it presumed that moral norm development could be stimulated within a legal setting. Therefore, the chapter concludes that a two-track approach is better suited to ensuring an ongoing process of development of both legal and moral norms.

A two-track approach in which legal and moral norm development are dealt with separately avoids the restriction of moral development by the legalistic context and can, thus, overcome the flaws of the ideal-typical model. These two tracks do not involve a strict separation of moral and legal norm development. Interaction between them can still contribute to a better adaptation of legal norms to practice, and improve their responsive character. Interaction between the two tracks should still be stimulated, as dynamics and interaction remain normative ideals for the interactive legislative approach. However, the development of one process should not depend on that of the other.

Section 10.2 explains the ethos of controversies. The section begins by exploring Waldron's ideas about the circumstances of politics and the difficulties posed by intractable moral disagreements. It contrasts Waldron's theory with that of Rawls, arguing that Rawls fails to adequately deal with conflicting opinions. It then goes further, linking Waldron's approach with that of agonists such as Mouffe, who focuses on the need to recognize conflict within democratic discussions. The section argues that Waldron and the agonists share with the ethos of controversies a respect for disagreement. Furthermore, it explains the three stages of an ethos of controversies.

Section 10.3 then explains why a two-track approach is required. It explores why it is counterproductive to aim for both legal and moral goals along one track. This section also discusses the conceptual distinction between legal and moral discourses. It argues that these discourses need to be distinguished in order to avoid the domination of moral norm development by a legalistic context. The interactive legislative approach should focus only on legal norm development and its legislative functions. Moral norm development should, then, be stimulated via other channels.

Section 10.4 discusses the first track of the two-track approach. It explains how an ethos of controversies can contribute to legal norm development. This section also clarifies the relation that should be maintained between the two tracks in order to stimulate the responsive character of the legal norms.

Section 10.5 discusses the second track of the two-track approach. It argues that moral norm development should be stimulated by methods outside the legal discourse. For example, ethics committees could play a prominent role in stimulating moral norm development. It goes on, however, to note that even with the inclusion of alternative tactics such as incorporating ethics committees, the moral track still face difficulties related to the strong tendency of the legal discourse to frame discussions in legalistic vocabulary. The section therefore concludes that further research is required to determine how to deal with the dominant legalistic context in which ethics committees are situated.

Finally, Section 10.6 presents some remaining remarks. It discusses the limits of an ethos of controversies for providing substantive justification of legal rules. Furthermore, it presents a refinement of one of the interactive approach's basic theses, as well as two new hypotheses for the ideal-typical model. These hypotheses will replace the claims concerning consensus and the interplay between moral and legal norm development. These hypotheses, it is argued, improve the practical value of the ideal-typical model of the interactive legislative approach.

10.2 An Ethos of Controversies

This section explains the ethos of controversies. An ethos of controversies can best be defined as a way of legal thinking that focuses on the controversies that characterize an issue. This ethos structures decision-making processes and norm development around an exploration of controversies, rather than a search for consensus. Additionally, the ethos of controversies emphasizes the political nature of decisions by pinpointing where conflicts may still exist.

An ethos of controversies responds to the failures of an ethos of consensus in addressing complex issues with significant moral impacts. For complex issues characterized by intractable moral disagreements, such as animal biotechnology, legal and moral norms have not yet crystallized. In other words, it seems that for these issues the current legal and ethical framework of vocabulary runs short.[2] The cognitive functions of law become important here. As Black argues, there is a translation problem, as the different disciplines involved all use different vocabulary. By providing a common vocabulary, law can help to solve these translation problems and integrate the various perspectives.[3] Sometimes, however, the reach of legal vocabulary runs short. Law, then, has to facilitate the development of new vocabulary. The best way of doing so, this section argues, is by leaving room for confrontation.

This section starts with explaining the circumstances of controversies. Explaining the circumstances will contribute to a better understanding of the remainder of this section: elaboration of an ethos of controversies.

10.2.1 The Circumstances of an Ethos of Controversies

To come to a better understanding of an ethos of controversies, we have to explain the circumstances it seeks to address. The section starts by elaborating Waldron's 'circumstances of politics', adapted from Rawls's discussion of the 'circumstances of justice'. The ethos of controversies corresponds to Waldron's ideas on circumstances of politics as both emphasize respect for disagreement. Although decisions have to be made, we need to be aware of the moral conflicts that may still exist. The ethos of controversies also shares the agonists' understanding of the conflicting nature of opinions in a pluralist democracy. Therefore, explaining the circumstances of an ethos of controversies also includes exploration of debate in agonic democracy.

2 Van Beers comes to similar conclusions related to human biotechnology. See B.C. van Beers, *Persoon en lichaam in het recht,* Den Haag, Boom Juridische Uitgevers, 2009, Ch. 2.

3 Black, 'Regulation as Facilitation'.

Waldron begins his argument by explaining Rawls's circumstances of justice. These circumstances are those aspects of the human condition that make justice necessary as a virtue and as a practice: moderate scarcity and the limited altruism of individuals.[4] According to Waldron, the circumstances of politics are developed along similar lines. There is a need for a common framework or course of action on some issues among members of a certain group. The need for common action, combined with a lack of agreement on what course of action to take or on the framework itself, define the circumstances of politics.[5] The common need for action brings about some difficulties. Individuals have different ideas and preferences about the best course of action, but they acknowledge that it is better to act together than to have no action or coordination at all.

Waldron argues that without these circumstances there would be no need to make political decisions. The need for a framework or course of action establishes a common need for political discourse. And without disagreement there would be no need for discussion in the first place.[6] Furthermore, the circumstances determine the authority of law. The authority of law depends not only on whether a particular decision can be identified as law, but also on whether decisions are respected as such. In order for people to respect the authority of law, Waldron argues, it is important to respect that disagreement may still exist. Respecting disagreement, even when a decision has been made, can do more justice to the various viewpoints of individual persons.[7] When disagreements are recognized and respected, it is more likely that individuals will be willing to respect the decision that has been made, even if it does not reflect their own preferences or viewpoints. Waldron therefore argues that the circumstances of politics require respect for disagreement. By the same token, a legal decision deserves respect if it addresses the circumstances of politics.

Although starting from a similar line of circumstances, Waldron's idea of democratic reasoning seems to differ from Rawls's. Rawls explains public reasoning by reference to a criterion of reciprocity, according to which citizens have the right to present claims and at the same time have the duty to give reasons for their claims if other citizens question them. Those reasons have to be based on convictions that already exist in public culture. Citizens, in this sense, have political autonomy. Whereas Waldron seeks to do justice to differing viewpoints by emphasising the role of disagreement in decision-making, Rawls focuses on the way in which citizens become willing to act in accordance with the principles of justice.[8] Rawls argues that in a liberal political regime, citizens reason together under fair terms of equality. Under these circumstances, democratic discussions will result in a reflective equilibrium that is grounded in 'settled convictions'. In sharing their public convictions, citizens will reach a consensus. Though individual citizens may still have opposing convictions and backgrounds, they will ultimately commit to shared values in the public sphere.

---

4 J. Rawls, *A Theory of Justice*, Cambridge, Harvard University Press, 1971, pp. 126-130.

5 J. Waldron, *Law and Disagreement*, Oxford, Clarendon Press, 1999, p. 102.

6 *Ibid.*, pp. 103-105.

7 *Ibid.*, p. 109.

8 Rawls, *A Theory of Justice*.

By confronting Rawls's conception of democratic debate and his understanding of a reflective equilibrium with Foucault's position on freedom, Rosenthal reveals the shortcomings of Rawls's consensus-orientated thinking. She argues that Rawls's reflective equilibrium results in a 'freestanding consensus' and risks excluding citizens who disagree.[9] Instead, she defends an ethics of democratic debate. In a nutshell, by fleshing out Foucault's position, she demonstrates that Rawls's conception of public reason privileges reasonableness over democratic contestation. Furthermore, he 'disregards the critical function of ontological claims in the politicization of dominant philosophical notions'.[10] Disregarding ontology and privileging reasonableness leads Rawls to underestimate the exclusionary effects of consensus. In doing so, he runs the risk that citizens will not be attached to ideals, which may even lead to antagonistic situations.

The agonistic alternative proposes an interplay between active citizenship and contested democracy. First of all, it underlines the importance of contesting both power relations and shared convictions. Second, agonistic democracy requires active citizenship in the sense that citizens must engage in agonic struggles. As a result, citizens must train their critical skills by engaging in an ongoing practice of critique. Social disharmony can also contribute to realizing active citizenship, as it can 'provide a starting-point for experimenting with ruling ideas and customs within this practice'.[11] Disharmony can stimulate an ongoing practice in which citizens can develop their skills and critiques. This development of critiques is an important feature of the ethos of controversies, as well.

Chantal Mouffe is a famous proponent of the agonistic tradition. According to Mouffe, deliberative democrats seem to ignore the conflicting dimension of pluralism, and erase the essential undecidability of politics with their strong focus on consensus. Mouffe argues that the deliberative idea of commonality is in fact politically constructed. She refers to deliberative democrats who perceive outcomes as being the result of moral consensus, and according to her, consequently fail to 'recognize the political nature of [their] own exclusion'.[12] In legal politics, decisions have to be made. At the same time, however, it is necessary to recognize the dimension of undecidability inherent in the conflicting nature of pluralism.[13] Recognizing conflict will therefore help to establish a legal discourse that is more realistic, and more just.[14]

This chapter argues that an ethos of controversies should be applied when using an interactive legislative approach to regulate issues with strong moral impacts. In particular, it argues that the intractable dimension of moral disagreements on issues such as animal biotechnology requires an ethos of controversies. Animal biotechnology issues are characterized by a strong moral pluralism and rapid changes. This moral disagreement requires

9 I. Rosenthal, *Democracy and Ontology: An Agonic Encounter between Political Liberalism, Foucault and Psychoanalysis,* dissertation, Amsterdam, VU University, 2012, p. 70, p. 115.

10 *Ibid.*, p. 114.

11 Rosenthal, *Democracy and Ontology,* p. 113.

12 C. Mouffe, 'Deliberative Democracy or Agonistic Pluralism', in *Social Research* 66/3, 1999, p. 756; A. Schaap, 'Agonism in Divided Societies', in *Philosophy & Social Criticism* 32/2, 2006, p. 262.

13 Mouffe, 'Deliberative Democracy or Agonistic Pluralism', p. 757.

14 *Ibid.*, p. 756.

further elaboration, as the various controversies that characterize the issue are still in development. Issues like these are difficult to address, because there is a current need for regulation despite strong (moral) disagreement. Such issues have a dimension of undecidability that cannot be resolved by the explication of clear-cut decisions. In addition, as demonstrated in the case-studies results and emphasized by the agonic democratic traditions discussed in this section, striving for consensus runs the risk of excluding citizens with conflicting positions. Thus, an ethos of controversies is better suited for stimulating an ongoing process of development.

The present research analyzes the assumption that controversial issues with strong moral impacts can be best addressed through an interactive legislative approach that aims for an ongoing process of norm development. This chapter adds that this process should follow an ethos of controversies. As stated in the beginning of this section, the ethos of controversies corresponds to Waldron's respect for disagreement; both are responses to similar political circumstances. Although decisions have to be made, we need to be aware of the moral conflicts that may still exist. As Waldron argues, this shows more respect for opinions that are excluded from decisions.

The ethos of controversies also shares the agonists' understanding of the conflicting nature of opinions in a pluralist democracy. However, an ethos of controversies maintains a strong deliberative dimension, as this ethos insists on a dialogue between actors. An ethos of controversies is a response to failures of an ethos of consensus, which the case studies showed to be insufficient for stimulating further norm development regarding complex issues with strong moral impacts. The presence of incommensurable values argues against consensus as a regulative ideal. But this does not mean that an ongoing process of norm development is impossible. As Rosenthal argues, the fact that there is a dimension of undecidability present in issues characterized by intractable moral disagreements does not mean that we should reject 'the aspiration for vibrant democratic discussions altogether'.[15] Even incommensurable values can be articulated and confronted with each other in a debate. An ethos of controversies values interaction, which helps to ensure that norms are orientated to practice. The articulation function of the interactive legislative approach facilitates the integration of the various viewpoints in during the decision-making process, if not in the final result.

10.2.2 An Ethos of Controversies

The ethos of controversies is designed to deal with issues that are characterized by controversial moral viewpoints. Instead of primarily aiming to bring controversial viewpoints closer together in order to come to an agreement, an ethos of controversies aims to do justice to the various viewpoints on an issue. An ethos of controversies presents a more realistic approach to legal decision-making processes, as it takes into account their political nature.

15 Rosenthal, *Democracy and Ontology.*

The design of an ethos of controversies differs from that of an ethos of consensus. An ethos of consensus structures deliberation through a common focus on consensus that can eventually legitimate decisions.[16] An ethos of controversies focuses instead on problem-definition and on structuring norm development. The ethos of controversies operates in three stages: articulation, confrontation, and awareness.

First of all, a focus on the controversies characterizing an issue means that the various viewpoints, concerns, and preferences must be *articulated*. A stocktaking of all these viewpoints, concerns, and preferences can contribute to problem-definition and may lead to a better understanding of the issue at stake. Instead of focusing on the commonalities among the various viewpoints, an ethos of controversies structures the discussion around a focus on the differences, thereby providing insights into the conflict. Without an emphasis on the controversies, there is a risk that viewpoints will be ignored,[17] disagreements will not be addressed, and problem definitions will be simplified. In order to avoid these risks, an ethos of controversies encourages the recognition of the conflicting viewpoints by showing respect to all of them.[18] A well-considered decision can only be made on the basis of a comprehensive description of the problem.[19] A problem-definition focused on conflict is thus a pre-condition for norm development and decision-making, as the process of problem-definition, here, aims to explicate the various viewpoints, concerns, and preferences of all parties. Without sufficient problem-definition, gaps in understanding the issues at stake will infect the solutions to the problem, as people begin from an incomplete definition of the problem.[20]

Second, debate in an ethos of controversies is built on a *confrontation* among these various viewpoints, concerns, and preferences. Debate is structured around a confrontation that acknowledges differences in reasoning. This forces the various actors to explain, to think through, but also to reconsider their viewpoints. Viewpoints, concerns and preferences are no longer loose statements, as their reasons are explicated and can be discussed. Confrontation, consequently, should contribute to stimulating further norm development.

Nevertheless, in the end, decisions have to be made.[21] It would undoubtedly be less complicated to make these decisions if everyone agreed. But, as the case studies make clear, issues such as animal biotechnology contain a dimension of undecidability. People express fundamental disagreements with respect to the moral impact of these technologies. Thus, controversies will still exist even after decisions are made. These controversies should not be excluded, but rather should be acknowledged in the decision-making

16 S. Benhabib, 'Toward a Deliberative Model of Democratic Legitimacy', in S. Benhabib (ed), *Democracy and Difference*, Princeton, NJ, Princeton University Press, 1996, pp. 67-94.

17 B. Barry, *Political Argument*, Berkeley, University of California Press, 1990, p. 240.

18 Waldron, *Law and Disagreement*, pp. 111-112.

19 Van Beers, Handelingen NJV 2009/I, p. 145.

20 J. Brand-Ballard, 'Consistency, Common Morality, and Reflective Equilibrium', in *Kennedy Institute of Ethics Journal*, 13/3, 2003, p. 255.

21 Bovenkerk, *The Biotechnology Debate*, p. 144.

process. The controversies may not be represented in the final decision, but they should appear in the reasoning that leads to these decisions. Exclusion of viewpoints cannot be prevented, but by continuing to explicate the controversies that define the problem, these positions will at least not become inferior.

The third stage of an ethos of controversies is therefore *awareness* of the conflict that still exists after decisions have been made. Awareness of conflict will contribute to acknowledging the political character of decisions. This acknowledgement in turn contributes to stimulating a more open debate during further norm development. This should encourage flexibility toward decisions that have been made, as these decisions will be confronted by conflicting viewpoints.

The interplay between these three stages stimulates an ongoing process of norm development. Furthermore, by operating in these three stages, an ethos of controversies aims to do justice to individual opinions. As a result, it is less likely that people will turn their back on the decisions that are made. Instead, they may be more willing to accept and respect decisions, even where those decisions may not reflect their own personal preferences and viewpoints.[22]

10.3 THE TWO-TRACK APPROACH

10.3.1 Why is a Two-Track Approach Required?

The previous section presented an ethos of controversies as an alternative to a focus on consensus in the interactive legislative approach. However, as mentioned before, an ethos of consensus is not the only explanation for the difficulties dynamic processes face. The case studies also showed that dynamic processes encountered difficulties related to the legalistic context in which they took place. Specifically, moral norm development stagnated under the influence of the legalistic framework that dominates practical thinking. The legalistic context contributed to the development of legal standards, which are required for decision-making. Moral norm development was dominated by the focus on developing legal standards. The strong tendency of the discourse to follow a legalistic framework seems to have been underestimated in theories of the interactive legislative approach. This tendency leaves less room for morality and the development of moral norms than the ideal-typical model assumed. Instead, the legalistic framework dominates legal politics and consequently moral norm development.

As described in Chapter 7, during the Dutch legislative process, legal and moral norm development strongly related to and indeed reinforced each other. The CAB ensured further interaction between these processes after implementation of the Law. The Dutch case seems to confirm the claims of Van der Burg and Brom concerning the reinforcement of legal and moral norm development. However, reinforcement was limited to the legislative

22 Rosenthal, *Democracy and Ontology*, p. 119.

process. Further moral norm development stagnated in legislative practice. The CAB gave advice on the moral acceptability of biotechnological procedures with animals. But because the CAB gave its advice during the licensing procedure, moral reflection and consequently also moral norm development could only take place in the context of this official procedure. Legal standards were developed during these procedures, but fundamental debates about the essence of the intrinsic value of animals were downplayed and stagnated prematurely. Consequently, the input for legal norm development stagnated as well.

In Switzerland, the legislature used a process-based approach for addressing animal biotechnology. As described in Chapter 6, however, the legislature still preferred the clear and concrete norms associated with traditional product-based legislative approaches. Because of this, the process-based approach ran the risk of collapsing into a product-based standard. And in reality it did: the fundamental basis of the concept of 'dignity of living beings' was not further challenged. Instead, 'dignity of living beings' was translated into clear standards that could be easily applied, excluding alternative conceptions.

Drawing on these outcomes, it seems that dynamics can only proceed so far in the context of the legal discourse. Although the Swiss and Dutch legislatures aimed to continue norm development, legal and moral norm development did not reinforce one another in these countries. An ethos of controversies may partly address these difficulties. This ethos can open up the legal discourse to create room for further norm development by emphasizing the political nature of decisions and recognizing controversies that still exist. Employing an ethos of controversies, then, creates room for incorporating morality in legal reasoning. This may stretch the legalistic framework of the legal discourse. It is questionable, however, whether an ethos of controversies can truly break through the legalistic framework required by basic elements such as legal security. Even though an ethos of controversies may create room for conflicting viewpoints, development is still restricted to the legal discourse. By following an ethos of controversies, law as such can articulate and translate different (moral) viewpoints into a commensurable framework, and thus give room for moral values. It is however, a framework of legal vocabulary and a discourse for legal norm development. Moral norm development has no explicit place in this discourse. Thus, by holding on to the assumption that moral norm development and legal norm development can be stimulated within the same discourse, moral norm development runs the risk of being limited by its legalistic context. Moral norm development as such therefore remains a problematic field with respect to the issue at stake.

To avoid those risks, an ethos of controversies should operate as part of a two-track approach for stimulating an ongoing process of moral and legal norm development. The two-track approach distinguishes moral and legal norm development from one another. Moral norm development may then have more room to follow a moral framework for reasoning. [23]

---

23 In order to better understand this approach, a remark must be made. These claims cannot be made for all kinds of issues. In the extent of this research, these claims can only count for issues such as animal biotechnology which are characterized by uncertainties, rapid developments and pervasive moral pluralism. Issues that have different circumstances may not require a two-track approach.

10.3.2 The Legal Discourse vs. the Moral Discourse

Legal discourse and moral discourse have various similarities, such as the concepts they use and the issues they address. However, each follows a different reasoning. It is important to be aware of the differences between legal and moral discourse and their methods of reasoning. Without awareness of the differences, without a distinction between these discourses, norm development may easily be caught in a single context, which will not ensure that norms are developed to their fullest extent. This section presents the main differences between legal and moral discourses and discusses the risks of failing to distinguish between them.

Distinction does not imply strict separation. Contrasting the legal/moral distinction with the functioning of self-referential closed systems exemplifies this.[24] In contrast to self-referential or autopoietic systems, legal discourse and values should interact and even intertwine with moral discourse and values. Moral discourse should not function as merely an external critique that requires a response only for the self-maintenance of the legal discourse.[25] Instead, moral discourse should interact with and shape law and politics. Taekema describes a shift from autonomous law to a responsive law that is more open to political values. These values are a response to social pressures and needs. A shift to responsive law makes law a *relatively* autonomous system.[26]

For complex problems with strong moral impact, the substantive moral dimension cannot be separated from the legal design, and thus involves an internal element as well. The diversity of viewpoints regarding issues with strong moral impacts complicates the legal issue and requires response. As a result, legal and moral discourses intertwine. For example, with respect to animal biotechnology, legal and moral discourses use similar concepts, such as animal integrity or dignity. At the same time, the reasoning of these discourses and the meaning of these concepts differ. Law facilitates integration of viewpoints by which it bridges the language problems. Law can translate the different vocabulary into a legal framework and by doing so integrate them. In this sense moral values do have a role. However, the role these values play in the legal framework differs from their position in moral discourse.

To begin with, legal and moral discourses frame issues differently. Similar to autopoietic systems, legal and moral discourses have their own methods of communication.[27] Legal discourse is based on legal institutions, whereas moral discourse is far more open-ended. Legal reasoning is framed by legal principles, legislation, the legal constitution, and international regulation. The development of legal norms has to fit with law, as explained in Chapter 2. Moral reasoning, on the other hand, is less constrained. Moral reasoning is mainly framed by normative ethical theories and moral codes.

---

24 G. Teubner, *Das Recht als Autopoietisches System*, Frankfurt am Main, Suhrkamp, 1989, p. 8.

25 *Ibid.*, Ch. 5; W. van der Burg, 'Law and Bioethics', in P. Singer and H. Kuhse (eds.), *A Companion to Bioethics*, Oxford, Blackwell, 2009, revised, second edition, pp. 56-64.

26 S. Taekema, *The Concept of Ideals in Legal Theory*, The Hague, Kluwer Law International, 2003, Ch. 6.

27 Teubner, *Das Recht als Autopoietisches System*, Ch. 5.

Furthermore, legal and moral discourses respond differently to external factors. For example, the political context is more influential for legal discourse than for moral discourse. Arguments of effectiveness or economic benefit, which are prominent in political discourse, are reflected in legal decision-making, but hardly influence moral reasoning. In addition, the translation of concepts in each discourse is different.

Nevertheless, there is a strong relation between these discourses, as questions concerning moral controversies and moral acceptability can influence the legal issue. The role of moral concepts in the legal framework is at least twofold. First of all, moral concepts can reflect public morality. In some cases, moral and legal discourses interact and even overlap. Moral concepts can sometimes be incorporated into legal discourse. This was the case, for example, with the regulation of euthanasia in the Netherlands. Van der Burg shows that for such matters, law and morality use similar approaches and the same conceptual categories.[28] In those cases, moral and legal norms can to a certain extent be developed along similar lines. Even in such cases, though, Van der Burg points out the risks of too much cooperation between legal and moral discourses.[29] For example, law and morality have different ideals regarding conceptual categories. He states that 'a too close co-operation may lead to neglecting important differences between law and morality. Law is an institution with a distinct role and distinct functions. This leads to specific possibilities and limitations of what law can and cannot achieve'.[30]

Second, in some cases, legal discourse 'borrows' moral concepts to reflect the moral impact of an issue. Where there is no unified public morality or shared framework about a moral issue, but instead a variety of conflicting moral values, moral concepts can function as 'contested concepts'[31] with aspirational value.[32] Contested concepts, here, are concepts about which the basic grounds as well as the meaning are still contested. Legal discourse uses these concepts to reflect that the issue has a strong moral impact that should be taken into account. Legal reasoning is thus able to avoid moral judgment and claim fairness in terms of neutrality, while in moral discourse, ethicists continue to search for answers and

---

28 Van der Burg, 'Law and Bioethics'.

29 Besides, we could question whether in the Dutch euthanasia debate the strong focus on consensus has not suppressed dissenting opinions. I will leave that aside as it goes beyond the context of this argument.

30 Van der Burg, 'Law and Bioethics', p. 71.

31 I derived 'contested concepts' from essentially contested concepts as was first coined by Gallie in 1956. Gallie developed a set of criteria to define these concepts. The criteria are identified as such: (I) their appraisive character, (II) internal complexity, (III) diverse describability, (IV) openness, (V)reciprocal recognition of their contested character among contending parties, (VI)an original exemplar that anchors conceptual meaning, and (VII) progressive competition, through which greater coherence of conceptual usage can be achieved. (see W.B. Gallie, 'Essentially Contested Concepts', in *Proceedings of the Aristotelian Society,* 1956, 56, pp. 167-198; D. Collier, F.D. Hidalgo, and A.O. Maciuceanu, 'Essentially Contested Concepts: Debates and Applications', in *Journal of Political Ideologies,* 2006, 11/3, pp. 211-246.) With the use of these concepts Gallie aimed to construct a coherent basis for discussing these complex concepts. However, the use of moral concepts in the context of this research does not fulfil all criteria. Besides, the aim of using these concepts does not follow from a theoretical perspective to construct a coherent basis for discussing these concepts. Instead, the use of contested concepts must be understood in legal politics. Therefore I choose to use a different concept which is merely inspired by Gallie's ideas about essentially contested concepts.

32 B. van Klink, *De wet als symbool,* p. 108.

moral judgment. Here, moral and legal norm development may follow very different lines. The contested concepts incorporated in the legal framework do not correspond with the moral concepts from which they originated. In these cases it is even more important to be aware of the differences between the two discourses as well as of the limits of dynamic interaction.

Failure to acknowledge the differences between legal and moral discourse can lead to neglecting the specific possibilities of both discourses as well as to a misunderstanding of the concepts used, leading to wrong expectations and connotations. It is therefore risky to stimulate legal and moral norm development within the same track. If moral norm development is captured by a legalistic framework, it may easily be mixed up with the development of legal standards that refer to similar moral concepts, as it is the legalistic framework that dominates practical thinking. This would be problematic, because while these legal standards may be sufficient to address the issues and their moral impact at the current time, they do not represent a conclusive statement on the moral impact of contested concepts. Besides, legal standards do not necessarily give meaning to moral concepts like animals' intrinsic value. The existence of legal standards concerning how to protect animals' intrinsic value, for example, does not mean that the concept of intrinsic value is no longer contested, either as a moral or as a fundamental legal concept.

10.4 A Track for Legal Norm Development

As the above argument has shown, a two-track approach that divides legal and moral norm development into separate processes is the most adequate way of avoiding the domination of the moral discussion by the legalistic context. This section takes a closer look at the first of these two tracks: the legal norm development track.

The legal norm development track reflects the process outlined in the ideal-typical model. Its main focus is on developing legal norms. In keeping with the interactive legislative approach, this track should contribute to developing the legal understanding of open norms. Changes in practice including the development of moral norms can influence the interpretation of legal norms. Therefore, interaction between practice, moral norm development and legal norm development should be encouraged and is considered one of the functions of law. However, this track differs from the ideal-typical model in that 1) it is restricted to the legal norm development process; and 2) it follows an ethos of controversies.

The legal track begins operating during the legislative process. During this phase the problem is defined, norms are developed, and the context for interpretation is given. As argued in Chapter 3, an important characteristic of the interactive legislative approach is that norm development continues after the law is drafted. The legislative process and the enactment of the law are thus only moments in the norm development process. Legislative practice can, for example, be organized as a licensing procedure. The licensing procedure is then the context in which further legal norm development takes place. Here, an ethos of controversies can play an important role as it ensures awareness of conflicts even after decisions have been made and legal standards have been partly developed. An ethos of controversies can build a bridge between legal norm development and any moral conflict that may still exist. This method of legal reasoning keeps legal norm development oriented to practice, and thus to the moral controversies as well.

Section 10.2.2 explained that the first phase of an ethos of controversies focuses on articulation: coming to understand the issue and its moral controversies. Second, an ethos of controversies can structure an ongoing process of norm development by promoting confrontation. Confrontation among the various norms, values, and intuitions both assists in norm development and improves their articulation. Third, awareness of controversies even after decision-making can ensure a process of continuous challenge and confrontation of a rule. As the case studies showed, if decisions are conceived as based on a consensus and lack awareness of the political nature of their exclusions, they run the risk of foreclosing further norm development. Conflicts can no longer contribute to a process of reflection and confrontation.[33] An ethos of controversies, by contrast, ensures awareness of opinions, values, norms, and intuitions that are excluded from decision-making, but which are, in fact, still part of the conflict. This encourages further norm development even after decisions are made.

Focusing on the controversies does not imply that coming to an agreement or setting legal standards at an early stage is not valuable. There is a need for controlling and guiding the development of these issues. Law can facilitate that. As Van der Burg claims, 'the most obvious institution for control is the law'.[34] Law is the institution for controlling and guiding the issue, and may be urgently required. Besides, constituting norms is a part of norm development. Both agreement and legal standards can contribute to a well-functioning norm development process. Without any doubt, some viewpoints must be excluded in order to actually come to a decision in practice. For example, some viewpoints may be less relevant for the particular case, or be overruled by other interests that seem of more importance at the current moment. Previous solutions and decisions influence new decisions, first and foremost due to the requirements of legal security. At the same time, if moments of decision-making were solely characterized by the application of previous solutions and decisions, there would be no room for further development of new norms. Decision-making should not necessarily imply that excluded viewpoints are no longer part of the general moral conflict. A dimension of undecidability will continue to exist for issues such as animal biotechnology. Consequently, further development is required. Stimulating further development presumes at least a certain level of dynamic interaction between the legal and moral norm development processes, even though the extent of this interaction is limited. An ethos of controversies may contribute to stretching the legal discourse's framework for interpretation and reasoning.

Awareness of the controversies, awareness of the exclusions from decision-making, and awareness of the undecidability of pluralism also stimulate comprehensive and fair reasoning in licensing and other processes of continued legal norm development. An ethos of controversies helps to establish a process in which legal standards and agreements can be challenged, further developed, and adapted to new developments in practice.

33 Bovenkerk and Poort, 'The Role of Ethics Committees in Public Debate', pp. 19-35.

34 Van der Burg, 'Law and Bioethics', p. 64.

10.5 A Track for Moral Norm Development

In the case-study countries, dealing with moral controversies and stimulating moral norm development proved particularly difficult. Moral norm development generally stagnated prematurely. Moreover, moral discourse was dominated by the legalistic context in which it had to operate. Therefore, the process of moral norm development should be stimulated in a moral track that can resist the dominance of the legalistic framework. In other words, the ideal-typical model of the interactive legislative approach should no longer hold on to the assumption that legal and moral norm development can be stimulated within the same context or along similar lines. The focus of the interactive legislative approach is legal norm development. However, in order to adjust to practice it must interact with the development of moral norms. These norms must, however, be developed within a different track: the moral track. This section will only briefly outline this moral track, as it is no longer part of the interactive legislative approach.

Moral debates can be stimulated by a call for regulation, as expert groups and advisory committees are directed to think through the moral impact of an issue. For some issues, traditional legal concepts offer sufficient vocabulary for complex moral debates. During the Dutch legislative process for regulating animal biotechnology, for example, legal and moral norms developed along similar lines. Nevertheless, as we have seen, moral debate differs from legal debate in that they use different framings and different conceptual ideals. The moral impact may characterize the legal issue, but the legal impact does not necessarily set the basis for the moral debate. Moral norm development must therefore be stimulated via a different track that is not bound by legal restrictions.

This moral norm development track could be structured in many ways. It could, for example, be characterized by the interplay between public deliberation and expert debates in which there is room for a moral framing of the issue. This combination can stimulate both a broadening and a deepening of input into norm development.[35] Public deliberation can provide insights into the concerns of members of society. Experts can translate people's concerns and viewpoints into reasonable moral arguments and provide the vocabulary for stimulating and structuring deliberation.

There are different examples of institutions that could be used to structure the moral track.[36] For example, ethics committees may contribute to stimulating moral norm development as well as to stimulating an interaction between moral discussions and legal norm development. At the same time, as the case studies show, the traditional means of institutionalising moral debate in ethics committees has some flaws. Also here, we should not underestimate the strong tendency to frame the debate in a legalistic context. Therefore, further research is required on how the moral track should be designed to avoid this domination.

---

35 Bovenkerk and Poort, 'The Role of Ethics Committees in Public Debate'.

36 See for example, A. Fung, 'Survey Article: Recipes for Public Spheres: Eight Institutional Design Choices and Their Consequences', in *The Journal of Political Philosophy*, 11/3, 2003, pp. 338-367.

10.6 Remaining Remarks

10.6.1 Critical Remarks

This chapter argues that the ethos of controversies and the two-track approach to legal and moral norm development are responses to the failures of the dynamic character of the ideal-typical model of the interactive legislative approach. However, some critical remarks must be made.

First of all, an ethos of controversies is a way of normative thinking in legal politics, and not a method for decision-making. Decisions have to be made, but these decisions require justification. As Somsen argues in the context of regulation on human biotechnology, due to the moral impact of these rules, justification must come from both a legal and a moral perspective.[37] An ethos of controversies cannot provide a substantive justification for legal norms, especially when operating along a two-track approach in which moral and legal norm development are separated. Nevertheless, an ethos of controversies can change ideas about reflection on critical morality as a source of substantive justification.

An ethos of controversies emphasizes the political character of legal decision-making, and does not necessarily resolve moral conflicts. Legal norms, thus, do not represent moral settlements. By emphasising that moral conflict still exists, and by explicating moral controversies, an ethos of controversies separates legal decision-making from moral judgement. A substantive justification in light of a critical moral reflection thus does not have to be based on a moral judgement. Neither should we search for substantive justification in any one normative moral theory. Instead, we should not adopt either one of these theories, but leave room for various methods of moral reasoning.[38]

Second, this discussion of the ethos of controversies and the two-track approach is limited to addressing particular types of issues. These ideas must be understood as addressing intractable moral disagreements such as animal biotechnology issues. This chapter presented modest claims about their extent, since it could not justify or argue for their adequacy for other issues. It may not, for example, provide an answer to issues such as multiculturalism. The Swiss case study pinpointed the blind spot of direct democracy, which risks excluding new minorities that do not have the capacity or the possibility to make use of their popular rights. Perhaps the interactive legislative approach risks similar blind spots, even with the use of an ethos of controversies.

The interactive legislative approach aims to co-operate on a horizontal level with all actors involved. In reality, however, it is doubtful whether all interested actors will actually be involved. Especially for issues characterized by cultural differences, it is questionable whether the parties involved will be willing to cooperate, or even have the capacity to do so. New minorities are not always organized or represented in interest groups. This can be addressed by stimulating involvement via other types of participatory methods. However, that does not ensure the willingness of minorities to participate. An ethos of controversies

37 Somsen, Handelingen NJV 2009/I, pp. 7-60.

38 Van Beers, Handelingen NJV 2009/I, pp. 99-145.

does not ensure minority input; it only makes the legal discourse more open to controversial viewpoints. To what extent are interaction and dynamics normative ideals for issues such as these? Or, to follow Rescher's terminology, are interaction and dynamics idealizations? Examining the application of the ideal-typical model to such issues would help to identify the limits of an ethos of controversies, and could be an interesting follow-up to this research.

10.6.2 Two New Hypotheses

This section presents two alternative hypotheses to complement the ideal-typical model of the interactive legislative approach. Furthermore, it suggests a refinement of one of the basic theses. These normative hypotheses replace the assumptions about the role of consensus and the dynamic interaction of moral and legal norm development, as these proved counterproductive in practice.

To begin with, the first basic thesis of the model should be refined by excluding the presumption that legal and moral norm development will always be mutually reinforcing. This assumption seems to ignore the reality of the limits of the legal discourse. The first basic thesis should therefore read:

The process of legislation on moral issues is structured as a process of interaction between the legislature and society or relevant sectors of society, which supports existing co-operation.

Second, two hypotheses should be added to the model. The first hypothesis replaces consensus as a regulative ideal with an ethos of controversies. It describes how, in the refined ideal-typical model of the interactive legislative approach, decision-making processes are structured:

In fields characterized by intractable moral disagreements, the interactive legislative approach follows an ethos of controversies. The articulation, challenge and awareness of conflicting viewpoints, concerns, and preferences during the decision-making process as well as after legal decision-making stimulates the dynamic process of ongoing norm development.

The second new hypothesis relates to the establishment of a two-track approach for moral and legal norm development. It describes how, in the refined ideal-typical model of the interactive legislative approach, norm development is structured:

In fields characterized by intractable moral disagreements, norm development occurs along two tracks that interact to ensure the ongoing orientation of legal norms on practice. One track focuses mainly on the ongoing process of legal norm development. The other track aims for an ongoing moral norm development process.

The amendment of the first basic thesis and the addition of these two new hypotheses will improve the practical value of the ideal-typical model.

Bibliography

Achen, T., *Den bioetiske udfordring. Et retspolitik studie af forholdet mellem etik, politik og ret I det lovforberedende arbejde vedrørende bio- og genteknologi i Danmark, Norge og Sverige*, Linköping, 1997.

Ammann D., and A.F. Goetschel, SAG-Studienpapiere, *Die Verfassungsnorm der Würde der Kreatur/Konsequenzen für die Zulassung genmanipulierter Tiere*, 1998.

Andersen, I., and B. Jaeger, 'Scenario Workshops and Consensus Conferences: Towards more Democratic Decision-making', in *Science and Public Policy*, 26/5, October 1999, pp. 331-340.

Andeweg, R.B., and G.A. Irwin, *Governance and Politics of the Netherlands*, Basingstoke, Palgrave MacMillan, 2002.

Asser-Scholten, *Algemeen Deel*, Zwolle, W.E.J. Tjeenk Willink, 1974 (3rd reprint).

Bakvis, H., 'Towards a Political Economy of Consociationalism. A Commentary on Marxist Views of Pillarization in the Netherlands', in *Comparative Politics*, 16/3, 1984, pp. 315-334.

Balzer, P., K.P. Rippe, and P. Schaber, *Menschenwürde vs Würde der Kreatur: Begriffsbestimmung, Gentechnik, Ethikkommissionen*, Freiburg/München, Alber Verlag, 1998.

Balzer, P., K.P. Rippe, and P. Schaber, 'Two Concepts of Dignity for Humans and Non-Human Organisms in the Context of Genetic Engineering', in *Journal of Agricultural and Environmental Ethics*, 13, 2000, pp. 7-27.

Barry, B., *Political Argument*, Berkeley, University of California Press, 1990.

Basse, E.M., and J. Dallberg-Larsen, 'The Danish Legal System', in H.T. Anker, B.E. Olsen, and A. Rønne (eds.), *Legal Systems and Wind Energy. A Comparative Perspective*, Alphen a/d Rijn, Kluwer Law International, 2008, pp. 56-85.

Benhabib, S., 'Toward a Deliberative Model of Democratic Legitimacy', in S. Benhabib, *Democracy and Difference*, Princeton, NJ: Princeton University Press, 1996, pp. 67-94.

Black, J. 'Regulation as Facilitation', *The Modern Law Review Limited* 1998, pp. 621-660.

Bos, D., M. Ebben, and H. te Velde (eds.), *Harmonie in Holland*, Amsterdam, Bert Bakker, 2007.

Bovenkerk B., and L.M. Poort, 'The Role of Ethics Committees in Public Debate', in *International Journal of Applied Philosophy*, 22/1, Spring 2008, pp. 19-35.

Bovenkerk, B., *The Biotechnology Debate. Democracy in the Face of Intractable Disagreement*, Dordrecht, Springer, 2011.

Brand-Ballard, J., 'Consistency, Common Morality, and Reflective Equilibrium', in *Kennedy Institute of Ethics Journal*, 13/3, 2003, pp. 231-258.

Brom, F.W.A., M.T. Hilhorst, R.H.J. ter Meulen, and J.M.G. Vorstenbosch, *Het toetsen van biotechnologische handelingen bij dieren. Rapport van een commissie van externe deskundigen ten behoeve van de Commissie Biotechnologie bij Dieren,* 1996.

Brom, F.W.A., *Onherstelbaar verbeterd. Biotechnologie bij dieren als een moreel probleem,* Assen, Van Gorcum, 1997.

Brom, F.W.A., 'Dynamische publieke moraal. Idealen als perspectief voor moraalontwikkeling', in W. van der Burg and F.W.A. Brom (eds.), *Over Idealen,* Deventer, W.E.J. Tjeenk Willink, 1998, pp. 75-90.

Brom, F., 'The Expressive-communication Function of the Bio-patent Legislation: The Need for Further Public Debate', in C. Baumgartner and D. Mieth (eds.), *Patente am Leben?,* Mentis, 2003.

Brunnee, J., and S.J. Toope, 'International Law and Constructivism: Elements of an Interactional Theory of International Law', in *Columbia Journal of Transnational Law,* 39/19, 2000-2001, pp. 19-74.

Brunnee, J., and S.J. Toope, *Legitimacy and Legality in International Law: An International Account,* Cambridge, Cambridge University Press, 2010.

Carlen, L., *Rechtsgeschichte der Schweiz,* Bern, Francke Verlag, 1988.

Church, C.H., *The Politics and Government of Switzerland,* Ltd. Houndmills Basingstoke Hampshire, Palgrave Macmillan, 2004.

Coglianese, C., 'Does Consensus Work? A Pragmatic Approach to Public Participation in the Regulatory Process', in Alfonso Morales (ed.), *Renascent Pragmatism. Studies in Law and Social Science,* Aldershot, Ashgate, 2003, pp. 180-195.

Cohen, J., 'Deliberation and Democratic Legitimacy', in J. Bohman and W. Rehg (eds.), *Deliberative Democracy, Essays on Reason and Politics,* Cambridge, MA, The MIT Press, 1997, pp. 242-277.

Collier, D., F.D. Hidalgo, and A.O. Maciuceanu, 'Essentially Contested Concepts: Debates and Applications', in *Journal of Political Ideologies,* 2006, 11/3, pp. 211-246.

Danish Council of Ethics, *Mand eller Mus,* 2007.

De Waard, W.N. (ed.), *Negotiated Decision-Making,* Den Haag, Boom Juridische Uitgevers, 2000.

Det Danske Selskab, *Nordic Democracy, Ideas, Issues, and Institutions in Politics, Economy, Education, Social and Cultural Affairs of Denmark, Finland, Iceland, Norway and Sweden,* Copenhagen, 1981.

Dorbeck-Jung, B.R., 'Een theoretisch kader voor wetgevingstoetsing. Op weg naar een responsieve wetgevingsleer', in *RegelMaat,* 4, 1998, pp. 198-209.

Dworkin, R., *Taking Rights Seriously,* Harvard, Harvard University Press, 1978.

Dworkin, R., *Law's Empire,* Harvard, Harvard University Press, 1993.

Eijlander, Ph. and W. Voermans, *Wetgevingsleer,* Den Haag, Boom Juridische Uitgevers, 2000.

Einsiedel. E., and D. Eastlick, 'Consensus Conferences as Deliberative Democracy: A Communications Perspective', in *Science Communication,* 21/4, June 2000, pp. 323-343.

Einsiedel, E., 'Publics at the Technology Table: The Consensus Conference in Denmark, Canada and Australia', in *Public Understanding of Science,* 10, 2001, pp. 83-98.

Einsiedel, E., 'Citizen Voices: Public Participation on Biotechnology', in *Notizie di Politeia,* 2001, pp. 1-15.

Errass, C., *Offentliches Recht der Gentechnologie im Ausserhumanbereich,* Bern, Stämpfli Verlag AG Bern, 2006.

Errass, C., 'Zum Verhältnis von Recht und Ethik in der Verfassungsbestimmung über die Gentechnologie im Ausserhumanerbereich', in *ZSR,* 2002/I.

Ethics Committee for Non Human gene technology (ECNH), and Swiss Committee on Animal Experiment (SCAE), a Joint Statement, *Dignity of Animals,* 2001.

Ethics Committee for Non Human gene technology (ECNH), and Swiss Committee on Animal Experiment (SCAE), a Joint Statement, *Research on Primates – An Ethical Evaluation,* 2006.

Fearon, J.D., 'Deliberation as Discussion', in J. Elster (ed.), *Deliberative Democracy,* Cambridge, Cambridge University Press, pp. 140-160.

Feld, L.P. and G. Kirchgassner, 'Direct Democracy, Political Culture, and the Outcome of Economic Policy: A Report on the Swiss Experience', in *European Journal of Political Economy,* 16, 2000, pp. 287-306.

Fleiner, T., 'Recent Developments of Swiss Federalism', in *The Journal of Federalism,* 32/2, 2002, pp. 97-124.

Florijn, N., 'Open wetgevingsproces, communicatieve wet? De totstandkoming van de Electriciteitswet 1998', in B. van Klink and W. Witteveen (eds.), *De overtuigende wetgever,* Deventer, W.E.J. Tjeenk Willink, 2000, pp. 211-224.

Fretz, L., and L. Sterrenburg (ed. A. Rijnen), 'Het maakbare dier geregeld: Politieke besluitvorming en genetische modificatie van dieren, een rapportage aan het Parlement', Den Haag, NOTA, 1993.

Fuller, L.L., *The Morality of Law,* New Haven, London, Yale University Press, 1964.

Fuller, L.L., *The Principles of Social Order,* Selected Essays of Lon L. Fuller, edited with an introduction by K. I Winston, Oxford, Hart Publishing, 2001.

Fung, A., 'Survey Article. Recipes for Public Spheres: Eight Institutional Design Choices and Their Consequences', in *The Journal of Political Philosophy,* 11/3, 2003, pp. 338-367.

Gallie, W.B., 'Essentially Contested Concepts', in *Proceedings of the Aristotelian Society,* 56, 1956, pp. 167-198.

Gamborg, C., M. Gjerris, J. Gunning, M. Hartlev, G. Meyer, P. Sandøe, G. Tveit, *Regulating Farm Animal Cloning: Recommendation of the Project*, Danish Centre for Bioethics and Risk Assessment, 2006.

Gould, C., 'Diversity and Democracy: Representing Differences', in S. Benhabib (ed.), *Democracy and Difference. Contesting the Boundaries of the Political*, Princeton, Princeton University Press, 1996, pp. 171-186.

Grotefeld, S., 'Self-Restraint and the Principle of Consent', in *Ethical Theory and Moral Practice*, 2000, pp. 77-92.

Grotefeld, S., 'Wie wird Moral ins Recht gesetzt?', in *Archiv. für Rechts- und Sozialphilosophie*, 2003, pp. 299-317.

Gutmann A., and D. Thompson, *Democracy and Disagreement*, Cambridge Mass., Harvard University Press, 1996.

Habermas, J., *Structural Transformation of the Public Sphere: An Inquiry into a Category of Bourgeois Society*, trans. by T. Burger and F. Lawrence, Cambridge, MA, The MIT Press, 1989.

Habermas, J., 'Three Normative Models of Democracy', in S. Benhabib (ed.), *Democracy and Difference. Contesting the Boundaries of the Political*, Princeton, NJ: Princeton University Press, 1996, pp. 21-30.

Hartlev, M., and U. Lind, 'Use of Blood Samples from the Danish PKU-Biobank. A Study of the Conceptualization of Research in Law and in Action', Paper for a workshop of EURECA, Tenerife, November 2005, pp. 1-28.

Hendriks, F., *Vitale democratie, theorie van democratie in actie*, Amsterdam, Amsterdam University Press, 2006.

Hertogh, M., 'De wondere wereld van de wetgever', in B. van Klink and W. Witteveen (eds.), *De overtuigende wetgever*, Deventer, W.E.J. Tjeenk Willink, 2000, pp. 45-60.

Hertogh, M., 'Deal and Dilemmas in Dutch Governance: Negotiated Decision-Making in the Netherlands', in B. De Waard (ed.), *Negotiated Decision-Making*, Den Haag, Boom Juridische Uitgevers, 2000, pp. 11-24.

Hisschemöller, M., and R. Hoppe, 'Coping with Intractable Controversies: The Case for Problem Structuring in Policy Design and Analysis', in *The International Journal of Knowledge Transfer and Utilization*, 1995, pp. 40-60.

Honig, B., *Political Theory and The Displacement of Politics*, Ithaca, NY, Cornell University Press, 1993.

Honoré, T., 'The Necessary Connection between Law and Morality', in *Oxford Legal Studies, 22*, 2002, pp. 489-495.

Joss S., and J. Durant, 'Introduction', in S. Joss and J. Durant (eds.), *Public Participation in Science. The Role of Consensus Conferences in Europe*, London, Science Museum, 1995, pp. 9-13.

Joss, S., 'Public Participation in Science and Technology Policy- and Decision-Making – Empheral Phenomen or Lasting Change?', in *Science and Public Policy,* 1999, pp. 290-293.

Kennedy, J., *Een weloverwogen dood. Euthanasie in Nederland*, Amsterdam, Prometheus, 2002.

Koh, H., 'Transnational Legal Process', in *Nebraska Law Review,* 75/181, 1996, pp. 181-206.

Klüver, L., 'Consensus Conferences in the Danish Board of Technology', in S. Joss and J. Durant (eds.), *Public Participation in Science. The Role of Consensus Conferences in Europe*, London, Science Museum, 1995, pp. 41-49.

Klüver, L., *EUROPTA. European Participatory Technology Assessment*, The Danish Board of Technology, 2000.

Lehnbruch, G., 'Consociational Democracy and Corporatism in Switzerland', in *Publius: The Journal of Federalism*, 23, Spring 1993, pp. 43-60.

Lesaffer, R., 'A Short Legal History of the Netherlands', in S. Taekema (ed.), *Understanding Dutch Law,* Den Haag, Boom Juridische uitgevers, 2004, pp. 31-58.

Lijphart, A., *The Politics of Accommodation: Pluralism and Democracy in the Netherlands,* Berkeley, CA, University of California Press, 1975.

Lijphart, A., and M.M.L. Crepaz, 'Corporatism and Consensus Democracy in Eighteen Countries: Conceptual and Empirical Linkages', in *British Journal of Political Science,* 21/2, 1991, pp. 235-246.

Lindahl, H., 'Communicatieve wetgeving en volkssoevereiniteit', in B. van Klink and W. Witteveen (eds.), *De overtuigende wetgever*, Deventer, W.E.J. Tjeenk Willink, 2000, pp. 173-188.

Linder, W., *Swiss Democracy: Possible Solutions to Conflict in Multicultural Societies*, New York, St. Martin's Press, 1994.

Linder, W., 'Direct Democracy', in U. Klöti, P. Knoepfel, H. Kriesi, W. Linder, Y. Papadopoulos, and P. Sciarini (eds.), *Handbook of Swiss Politics*, Zürich: Neue Zürcher Zeitung, 2007.

Linskens, M., *Transgene landbouwhuisdieren: Het overwegen waard?,* Wageningen, Nederlandse Christelijke Plattelandsjongeren/Wetenschapwinkel LUW, rapport 31, 1989.

March, J.G., and J.P. Olsen, *Rediscovering Institutions: The Organizational Basis for Politics*, New York, The Free Press, 1989.

Meijer, A., G.K. Pikker, M.J.W.A. Schiffelers, and A.M.J. van der Spek., *Evaluatie van het Besluit Biotechnologie bij Dieren*, 2005.

Meijer, A.J., and F.W.A. Brom, 'Biotechnology and Social Learning: An Empirical Analysis of the Dutch Animal Biotechnology Act', in *Technology in Society,* 2008.

Moreno, J.D., 'Biotech is thrusting us into new political territory', *newscientist.com Issue 2879*, 28 August 2012.

Mouffe, C., 'Deliberative Democracy or Agonistic Pluralism', in *Social Research*, 66/3, 1999, pp. 745-758.

Nonet, P., and P. Selznick, *Law and Society in Transition: Toward Responsive Law*, New York, Harper & Row, 1978.

Paula, L.E., 'Biotechnologie bij dieren ethisch getoetst? Een onderzoek naar het functioneren van het Besluit Biotechnologie bij Dieren', Den Haag, Rathenau Insituut, 2001.

Paula, L.E., *Ethics Committees, Public Debate and Regulation: An Evaluation of Policy Instrument in Bioethics Governance*, Amsterdam, Athena Institute, 2008.

Perrez, F.X., C. Errass, and K. Bende, 'GMO-regulation: Case Study of Switzerland', December 2003, <www.law.nyu.edu>, pp. 3-48.

Porsborg Nielsen, A., J. Lassen, and P. Sandøe, 'Democracy at its Best? The Consensus Conference in a Cross-National Perspective', in *Journal of Agricultural and Environmental Ethics*, 20, 2007, pp. 13-35.

Porsborg Nielsen, A., J. Lassen, and P. Sandøe, 'Involving the Public. Participatory Methods and Democratic Ideals', <bioethics.kvl.dk>.

Rawls, J., *A Theory of Justice*, Cambridge, Harvard University Press, 1971.

Rawls, J., *Political Liberalism*, New York, Columbia University Press, 1996.

Reich, J., 'The Interactionist Model of Direct Democracy. Lessons from the Swiss Experiences', SSRN, 2008, pp. 16-31.

Rescher, N., *Pluralism, Against the Demand for Consensus*, Oxford, Clarendon Press, 1993.

Rendtorff, J.D., and P. Kemp, *Basic Ethical Principles in European Bioethics and Biolaw, Vol. 1 Autonomy, Dignity, Integrity and Vulnerability*, Barcelona, Impremta Balona, 2000.

Renn, O., T. Webler, H. Rakel, P. Dienel, and B. Johnson, 'Public Participation in Decision Making: A Three-Step Procedure', in *Policy Sciences*, 26, 1993, pp. 189-214.

Rippe, K.P., 'Ethikkommissionen als Expertengremien? Das Beispiel der Edgenössischen Ethikkommission', in K.P. Rippe (ed.), *Angewandte Ethik in der pluralistischen Gesellschaft*, Fribourg, Freiburger Universitätsverlag, 1999, pp. 359-370.

Rochon, T.R., *The Netherlands. Negotiating Sovereignty in and Independent World*, Boulder, Co, Westview Press, 1999.

Rosenthal, I., *Democracy and Ontology: An Agonic Encounter between Political Liberalism, Foucault and Psychoanalysis*, dissertation, Amsterdam, VU University, 2012.

Ross, A., *On Law and Justice*, The Lawbook Exchange, Ltd, Clark New Jersey, 2004 (reprint).

Ross, A., 'Livets Hellighed contra individets autonomi', in *Ugeskrift for læger*, Copenhagen, 1979.

Rothmar Hermann, J., and M. Rowlandson, 'The Role of Ethics and Morality in EU Law', in *Journal of International Biotechnology Law*, 5, 2008, pp. 241-244.

Rothmar, Hermann, J., 'Are you a Man or a Mouse?', in *Journal of International Biotechnology Law,* 2008, pp. 16-19.

Schaap, A., 'Agonism in Divided Societies', in *Philosophy & Social Criticism*, 32/2, 2006, pp. 255-277.

Schellekens, H., *De DNA-Makers*, 1993.

Secretariat of the Convention on Biological Diversity, 'Cartagena Protocol on Biosafety, Article 3 (I)', 2000.

Selznick, P., *The Moral Commonwealth. Social Theory and the Promise of Community*, Berkeley, CA, University of California, 1992.

Somsen, H., Handelingen NJV 2009/I, pp. 7-60.

Stamhuis, J., *Conflicting Interests in Corporate Regulation*, Enschede, Gildeprint Drukkerijen, 2006.

Stamhuis, J., 'Communicative Law: A Quest for Consensus', in N. Zeegers, W. Witteveen, and B. van Klink (eds.), *Social and Symbolic effects of Legislation under the Rule of Law,* Lampeter, Ceredigion, Wales, The Edwin Mellen Press, 2005, pp. 277-298.

Stichting van de Arbeid, 'Centrale aanbevelingen inzake aspecten van een werkgelegenheidsbeleid' (Akkoord Van Wassenaar), 1982.

Stout, H.D., *De betekenissen van de wet. Theoretisch-kritische beschouwingen over het principe van wetmatigheid van bestuur*, Zwolle, W.E.J. Tjeenk Willink, 1994.

Sunstein, C.R., *Legal Reasoning and Political Conflict,* New York, Oxford University Press, 1996.

Szladits, C., *A Guide to Foreign Legal Materials*, Dobbs Ferry, Oceana Publications Inc., 1959.

Taekema, H.S., *The Concept of Ideals in Legal Theory,* The Hague, Kluwer Law International, 2003.

Taekema, S., *Understanding Dutch Law,* Den Haag, Boom Juridische uitgevers, 2004.

Tamm, D., 'The Danes and Their Legal Heritage', in B. Dahl, T. Melchior, and D. Tamm (eds.), *Danish Law in European Perspective*, Copenhagen, Thomson Publishers, 2002, pp. 41-60.

Teknologinævnet (Danish Board of Technology), *Teknologiske dyr, Indgreb I kønsceller på højerestående dyr.* Slotdokument og ekspertoplæg fra konsensuskonferencen 23-25 September 1992, arrangneret af Tekologiinævnet i samarbejde med Folketingets Forskningsudvalg, København 1993.

Teubner, G., *Das Recht als Autopoietisches System*, Frankfurt am Main, Suhrkamp, 1989.

Tveit, G., K.H. Madsen, and P. Sandøe, *Biotechnology and the Public Debate*, Danish Centre for Bioethics and Risk-Assessment, 2003.

Van Beers, Handelingen NJV 2009/I, pp. 99-145.

Van Beers, B.C., *Persoon en lichaam in het recht,* Den Haag, Boom Juridische Uitgevers, 2009.

Van der Bruggen, K., 'Dolly and Polly in the Polder: Debating the Dutch Debate on Cloning', in *The Public Debate on Cloning International Experiences*, The Hague, Rathenau Institute, 1999.

Van der Burg, W., 'The Importance of Ideals', in *The Journal of Value Inquiry*, 31, 1997, pp. 23-37.

Van der Burg, W., 'Wetgeving inzake morele kwesties: Een procesgerichte benadering', in *Ars Aequi*, 47/2, 1998, pp. 79-87.

Van der Burg, W., 'Morality of Aspiration', in W. van der Burg and W. Witteveen (eds.), *Rediscovering Fuller: Essays on Implicit Law and Constitutional Design,* Amsterdam, Amsterdam University Press, 1999, pp. 169-192.

Van der Burg, W., and F.W.A. Brom, 'Legislation on Ethical Issues: Towards an Interactive Paradigm', in *Ethical Theory and Moral Practice,* 3/1, 2000, pp. 57-75.

Van der Burg, W., 'The Expressive and Communicative Functions of Law, Especially with Regard to Moral Issues', in *Law and Philosophy,* 20/1, 2001, pp. 31-59.

Van der Burg, W., 'The Irony of a Symbolic Crusade: The Debate on Opening Up Civil Marriage for Same-Sex Couples', in N. Zeegers, W. Witteveen, and B. van Klink (eds.), *Social and Symbolic Effects of Legislation Under the Rule of Law,* Lewiston New York, The Edwin Mellen Press, 2005.

Van der Burg, W., Het ideaal van de neutrale staat. Inclusieve, exclusieve en compenserende visies op godsdienst en cultuur, Inaugural Lecture, Erasmus University, Rotterdam, 2008.

Van der Burg, W., 'Law and Bioethics', in P. Singer and H. Kuhse (eds.), *A Companion to Bioethics*, Oxford, Blackwell Companion Guide, 2009, pp. 49-57.

Van Klink, B.M.J., *De wet als symbool. Over wettelijke communicatie en de Wet gelijke behandeling van mannen en vrouwen bij de arbeid,* Deventer, W.E.J. Tjeenk Willink, 1998.

Van Klink, B., and W. Witteveen, 'Why Is Soft Law Really Law?', in *RegelMaat*, 3, 1999, pp. 120-140.

Van Klink, B., 'Legislation in the Netherlands: The Communicative Turn', in S. Taekema (ed.), *Understanding Dutch Law,* Den Haag, Boom Juridische uitgevers, 2004, pp. 103-129.

Van Manen, N.F., 'Rechterlijke rechtsvinding en intuïtie', in E.A. Lissenberg, P.W. Brouwer, F. Jacobs, and E. Poortinga (eds.), *De actualiteit van Paul Scholten,* Nijmegen, Ars Aequi Libri, 1996, pp. 57-74.

Van Schooten, H., 'Instrumental Legislation and Communication Theories', in H. van Schooten (ed.), *Semiotics and Legislation: Jurisprudential, Institutional and Sociological Perspectives*, Liverpool, Deborah Charles, 1999, pp. 185-211.

Van Schooten, H., 'Van bevel naar communicatie', in B. van Klink and W. Witteveen (eds.), *De overtuigende wetgever,* Deventer, W.E.J. Tjeenk Willink, 2000, pp. 71-86.

Van der Sluijs, J.P., R. van Elst, and M. Riphagen, 'Beyond Consensus: Reflection from a Democratic Perspective on the Interaction between Climate Politics and Science', in *Elsevier Editorial System for Current Opinion in Environmental Sustainability*, 2010/2, pp. 409-415.

Van der Vlies, I.C., *Het wetsbegrip en beginselen van behoorlijke wetgeving: Het legaliteitsbeginsel*, Den Haag, VYGA, 1984.

Vatter, A., 'Consensus and Direct Democracy: Conceptual and Empirical Linkages', in *European Journal of Political Research*, 38, 2000, pp. 171-192.

Vatter, A., 'Federalism', in U. Klöti, P. Knoepfel, H. Kriesi, W. Linder, Y. Papadopoulos, and P. Sciarini (eds.), *Handbook of Swiss Politics*, Zürich, Neue Zürcher Zeitung, 2007.

Visak, T., and F. Meijboom, 'Integriteit van dieren: Bouwsteen of Struikelblok? Verslag van de eerste discussiebijeenkomst over biotechnologie bij dieren', Utrecht, Centre for Bioethics and Health Law, 2002.

Visak, T., and F. Meijboom, 'Op zoek naar alternatieven – maar voor welke doelstelling?', Utrecht, Ethics Institute, 2003a.

Visak, T., and F. Meijboom, 'Het maatschappelijk belang van een doelstelling: Zoeken naar goede argumenten. Verslag van de derde discussiebijeenkomst over biotechnologie bij dieren', Utrecht, Ethics Institute, 2003b.

Visak, T., and F. Meijboom, 'De waarde van een machtsvrije dialoog voor verdere verdieping. Verslag van de vierde discussiebijeenkomst over biotechnologie bij dieren', Utrecht, Ethics Institute, 2004.

Voorlopige commissie ethische toetsing genetische modificatie van dieren (Schroten II), *Advies inzake het dossier 'Weefselspecifieke expressie van genen in de melkklier van genetisch gemodificeerde runderen'*, Den Haag/Utrecht, LNV, 1992.

Vorstenbosch, J., and P. Ippel, 'De weg naar de wet. Wetgeving over ethische kwesties: Een interactief proces', in W. van der Burg and P. Ippel (eds.), *De Siamese tweeling*, Assen, Van Gorcum, 1994, pp. 49-67.

Waldron, J., *Law and Disagreement*, Oxford, Clarendon Press, 1999.

Westerman, P., 'Means and Ends', in W. Van der Burg and W. Witteveen (eds.), *Rediscovering Fuller: Essays on Implicit Law and Constitutional Design*, Amsterdam, Amsterdam University Press, 1999, pp. 47-62.

Witteveen, W.J., P. van Seters, and G. van Roermund (eds.), *Wat maakt de wet symbolisch?*, Zwolle, W.E.J. Tjeenk Willink, 1991.

Witteveen, W., and B. van Klink, 'De overtuigende wetgever', in B. van Klink and W. Witteveen (eds.), *De overtuigende wetgever*, Deventer, W.E.J. Tjeenk Willink, 2000, pp. 5-28.

Witteveen, W., 'Turning to Communication in the Study of Law', in N. Zeegers, W. Witteveen, and B. van Klink (eds.), *Social and Symbolic Effects of Legislation under the Rule of Law*, Lampeter, Ceredigion, The Edwin Mellen Press, 2005.

Witteveen, W.J., 'A Self-Regulation Paradox: Notes towards the Social Logic of Regulation', *EJCL* 2005-1.

Witteveen, W., *Het wetgevend oordeel,* Den Haag, Boom Juridische Uitgevers, 2010.

Wolthuis, B.J., 'Rechtsvorming als debat: De overtuiging het juiste antwoord te geven als voorwaarde voor goede rechtsvorming', in E.J. Broers and B. van Klink (eds.), *De rechter als rechtsvormer*, Den Haag, Boom Juridische Uitgevers, pp. 108-112.

Parliamentary Documents

Denmark

Danish Ministry of Trade and Industry, *An Ethical Foundation for Genetic Engineering Choices*, Denmark, 1999.

Danish Ministry of Trade and Industry, *The Danish Government Statement on Ethics and Genetic Engineering*, August 2000.

Justitsminister, *Skriftlig Fremsættelse*, L25, Forslag til lov om cloning og genmodificering af dyr, 7 October 2004.

Justitsminister, *Kommentaret oversigt over høringssvar om forlsag til lov om kloning og genmodificering af dyr m.v.*, L8, 24 February 2005.

Ministeriet for Videnskab, Teknologi og Udvikling, *Genmodificerede og klonede dyr*, 15 October 2002.

Switzerland

Bericht über die Ergebnisse des Vernehmlassungsverfahrens zum Vorentwurf der Gen-Lex-Vorlage', Bern, 1998.

Eidgenössisches Volkswirtschaftsdepartement, 'Bericht über die Ergebnisse des Vernehmlassungsverfahrens zum Vorentwurf der Gen-Lex-Vorlage', Bern, 1998.

The Netherlands

Explanatory Memorandum, Animal Health Welfare Law, 1992.

Explanatory Memorandum, Animal Biotechnology Decree, 1997.

Letter of the Minister of Agriculture, Nature and Fishery tot the House Representatives, dd. 28 November 1991,19744, nr. 8.

Parliamentary Documents, Second Chamber, 1980-1981, 16447.

Parliamentary Documents, Second Chamber, 1988-1989, 16447.

Parliamentary Documents, Second Chamber, 1889-1990, 16447.

Parliamentary Documents, Second Chamber, 1995-1996, 24400 XIV.

Rapport van Tijdelijke Commissie van Ethiek en Biotechnologie bij Dieren 1990, NRLO/90/ W 55, Den Haag.

WODC, *Nota Zicht op Wetgeving*, 2003.

Appendix I List of Interviewees

Denmark

Prof. H.T. Anker — *October 2007*
Professor of Environmental Law, University of Copenhagen

Prof. M. Hartlev — *October 2007*
Member of the Danish Ethics Council
Professor of Healthcare Law and Bioethics, University of Copenhagen

U. Hybel — *October 2007*
Member of the Danish Ethics Council

Prof. P. Kemp — *December 2007*
Chair Centre for Ethics and Law
Director of Department of Philosophy of Education,
Danish University of Education, Copenhagen

Prof. A. Kornerup Hansen — *October 2007*
Member of the Animal Experiment Directorate
Professor of Vetinary Disease Biology, University of Copenhagen

L. Klüver — *October 2007*
Executive director Danish Board of Technology

G. Meyer — *November 2007*
Journalist, associated with Centre for Bioethics and Risk Assessment

H.P. Olessen — *October 2007*
Secretary Animal Experiment Directorate

Dr. J. Rendtorff — *November 2007*
PhD at Centre for Ethics and Law
Lector Organisations, Change and Management at Roskilde University

J. Rothmar Hermann — *November 2007*
PhD-candidate Healthcare Law, University of Copenhagen

Prof. P. Sandøe — *September, October, November 2007*
Chair of the Danish Council for Animal Ethics
Professor of Bioethics, Centre for Bioethics and Risk Assessment,
University of Copenhagen

Switzerland

Prof. Dr. K. Buerki *April 2007*
member of the Ethics Committee on Gene Technology in the Non-Human Field
Director of the Institute of Laboratory Animal Science, University of Zurich

Dr. Ch. Errass *March 2007*
Author of *Offentliches Recht der Gentechnologie im Ausserhumanbereich,* Bern, Stämpfli Verlag AG Bern, 2006.

H. Hochbach *March 2007*
Head of Section Biotechnology, Federal Office for the Environment

Prof. Dr. K.P. Rippe *February 2007*
Chair of the Ethics Committee on Gene Technology in the Non-Human Field
Professor of Philosophy at the University of Education, University of Karlsruhe

Prof. Dr. P. Schaber *January 2007*
Professor of Applied Ethics at the Ethics Research Institute, University of Zürich

Prof. Dr. M. Schefer *April 2007*
Member of the Ethics Committee on Gene Technology in the Non-Human Field
Professor of Constitutional and Administrative Law, University of Basel

A. Willemsen *January, March 2007*
Secretary of the Ethics Committee on Gene Technology in the Non-Human Field

The Netherlands

Prof. E. Schroten *June 2009*
Chair of the Committee on Animal Biotechnology

S. Beukema *June 2009*
Secretary of the Committee on Animal Biotechnology

Index

A

Accommodation, politics of, 97, 111–113
Agonic democracy, 153, 155
Agonism, 148, 151–156
Agreement, 102, 108, 112, 130, 145, 149, 163
  incompletely theorized agreements, 150
Animal biotechnology, 3–10, 12–14
  regulation of animal biotechnology, 49–50, 52–66, 72, 75, 81, 83, 94–96, 120, 122–123, 125–128, 130
Animals
  dignity of animals, 9, 13, 122, 160
  integrity of animals, 9, 14, 50–53, 72, 123, 125, 145, 160
  intrinsic value of animals, 98, 100–101, 107–108, 123, 127, 145, 159, 162
Approaches to law
  alternative approaches to law, 10
  process-based approach, 79, 94, 96, 133
  product-based approach, 79, 94, 96, 133
Authority of law, 11, 154
Autopoietic systems, 160

B

Balzer, P., 84
Bargaining, 62, 89, 92, 96
Barry, B., 157
Basse, E.M., 56, 59, 61, 64, 71–72
Benhabib, S., 142, 157
Biolaw, 69, 71
Biosafety, 7, 80
BioTIK, 57, 68–69
Black, J., 3, 5, 10–11, 28, 32, 36, 38, 153
Bovenkerk, B., 4, 8, 103–104, 106, 108, 112–113, 115–116, 149–150, 157, 163–164
Brand-Ballard, J., 145
Brom, F.W., 4, 7, 10–12, 22, 27–28, 31–35, 37, 99, 102, 106, 110, 139, 141
Brunnee, F.W., 6, 10–11
Brunnee, J., 28, 31, 33–35, 38–39

C

Church, C.H., 47
Circumstances
  circumstances of an ethos of controversies, 153, 155–157, 162
  circumstances of justice, 154
  circumstances of politics, 152–154
Co-operative effort, 30–31, 38, 119, 121, 132
Coglianese, C., 149
Cohen, J., 142
Communication, 3, 6, 10, 12–14, 27–31, 34–36, 38, 40, 62, 73, 96, 109, 117, 122, 129, 142, 160
Complex issues, 4, 6–10, 21, 133, 136, 153–156
Conflict, 6, 9–10, 71, 79, 86, 89, 108, 111, 113, 148–150, 162–163, 165–166
Confrontation, 151, 153, 157, 163
Consensual style of democracy, 87, 89, 95, 111–112, 114, 116–117, 131, 135
Consensus, 6, 9, 12, 15, 34–35, 38, 85–90, 103–104, 109, 112, 116–117, 125–128, 135, 139, 140–149, 154–155
  consensus as the ideal outcome, 139–143
  consensus as a regulative ideal, 37, 126, 134, 139, 140–142, 144, 146
  consensus conferences, 14, 49–50, 56, 62–63, 65, 120, 126–127

Consensus (*continued*)
ethos of consensus, 140, 146, 149, 156–157
Consociational style of democracy, 111
Contestation, 144, 148, 155
Contested concepts, 161
Controversies, 77, 107, 124, 126, 129–130, 134, 145, 147, 150–153, 155–159, 162–163, 165
circumstances of an ethos of controversies, 153, 155–156, 162
ethos of controversies, 15, 150–153, 155–159, 162–163, 165

D

Dallberg-Larsen, J. *See* Basse
Danish Animal Ethics Committee, 51, 57, 67–69
Danish Ethics Council, 64, 70
Debate
ongoing debate, 4–5, 10, 12–14, 17, 26, 33–34, 37–38, 40
public debate, 14, 47, 50, 56, 65, 67–68, 70, 82, 121–122, 125, 131, 135
Decision-making
guidelines for ethical decision-making 58
horizontal decision-making, 10, 12, 31, 34, 131–132
negotiated decision-making 113, 117
transparency of decision-making 56, 60, 64–65, 114
vertical decision-making 30
Deliberative democracy, 67, 72, 135, 139, 142, 155–156
Democracy
agonic democracy, 153, 155–156
blind spot of democracy, 90, 96
consensual style of democracy, 87, 89, 95, 111–112, 114, 116, 131, 135
deliberative democracy, 67, 72, 135, 139, 142, 155–156
direct democracy, 13, 86–90, 92, 94–95, 131–132
Denmark, 6, 14, 49–50, 53–56, 59–63, 70, 72–73
Animal Cloning Act, 49, 51, 53, 56–57, 59
BioTIK, 57, 68–69
legislative culture, 70, 72
political culture, 61, 63, 65
Dialogue, 29, 83, 141, 156
Dignity
dignity of animals, 9, 13, 122, 160
dignity of living beings, 76–78, 80–81, 83–86, 95–96, 120, 123–126, 128, 130, 145, 147, 159
Dimension
dimension of fit, 19, 21, 25
dimension of legal validity, 24, 26
dimension of political morality, 21–23, 25
Direct democracy, 13, 86–88, 90, 92, 94–95, 131–132
Disagreement, 3–4, 8–9, 85, 110, 113, 139, 153–157
intractable disagreements, 9–10, 21, 141, 144, 149, 155–156, 165–166
Durant, J., 65–66
Dutch Committee of Animal Biotechnology, 14, 100–106, 108, 118, 135
Dworkin, R., 17–19, 22–23, 26, 148
Dynamics, 11, 31, 33, 102, 119, 134–136, 147, 149–150, 158–159, 162–163, 165–166

E

Einsiedel, E., 64, 66
Errass, C., 76–79, 84, 94
Ethics committees, 13, 15, 134, 152, 164

F

Facilitation, regulation as, 5, 10, 28, 33, 36, 38, 153
Federalism, 86–88, 90
Flexibility, 9, 12, 28–29, 33, 117–118, 129–130, 158
Foucault, M., 155
Fuller, L.L., 6, 19, 21, 23, 27–31, 34, 39, 116
Functions of law
- articulation function of law, 35
- cognitive function of law, 6, 36
- communicative function of law, 6
- coordination function of law, 35
- expressive function of law, 87, 122, 128
- instrumental function of law, 73, 94, 117
- interaction function of law, 6, 35

G

Gallie, W.B., 161
Grotefeld, S., 32, 93–95, 148
Gutmann, A., 139, 142

H

Habermas, J., 139–140, 142–144
Hisschemöller, M., 21
Honig, B., 147
Honoré, R., 23
Hoppe, R. *See* Hisschemöller

I

Integrity of animals, 9, 14, 50–53, 72, 123, 125, 145, 160
Interaction, 27–36, 105, 116, 119, 126, 136, 139, 143, 152, 156, 158, 162–164, 166
Interactional international theory, 6, 10, 33
Interactionism, 4, 10–12, 21, 29–30, 116–118, 133–134, 136
Interactive legislative approach, 4–6, 10–12, 27–40, 43, 119, 121, 127, 129, 131–136, 139–141, 144, 146, 152, 155–156, 158, 162, 164–166
Intrinsic value of animals, 98, 100–101, 107–108, 123, 127, 145, 159, 162
Involvement
- expert involvement, 65–67, 72, 74, 93, 95, 114, 118, 120, 131
- public involvement, 14, 24, 33, 55–56, 65–66, 81–83, 88, 90, 93, 105–106, 114

Ippel, P., 4, 36

J

Joss, S. *See* Durant
Justification, 9
- substantive justification, 17, 21–23, 25, 127–128, 144–146, 152, 165

K

Kemp, P., 52, 68
Klüver, L., 63, 65, 67

L

Law
- integrity of law, 17
- pragmatic view of law, 6, 59, 68–69, 71, 73, 117, 133, 146
- quality of law, 6, 18–26, 31, 39, 127

Legal positivism, 25, 94–95, 116, 133, 135
Legality, practice of, 19, 31, 34–35, 39

Legitimacy
democratic legitimacy, 12, 24–25, 109, 114, 116
formal legitimacy, 24–26
political legitimacy, 5, 12, 32
popular legitimacy, 10, 17, 21–24, 26, 39–40, 59, 73, 94, 96, 109–110, 127–128, 131, 144
Lehnbruch, G., 89–90, 92
Licensing procedure, 13, 53–55, 76, 80, 101–103, 105–106, 120–125, 129–130, 132
Lijphart, A., 87, 89, 111–112, 114
Linder, W., 86–88, 90–91

M
March, J.G., 62,
Moral issues, 4–5, 51, 53, 60, 69, 72–73, 83, 93, 105, 124, 130, 133, 140
Moral values
in legislation, 107
in regulation, 44, 48, 55–56, 83, 135
Morality, 11, 94, 145, 161
critical morality, 11, 22–23, 25, 40, 144, 165
morality of aspiration, 30, 39
morality of duty, 30
political morality, 14, 21–23, 26
public morality, 22–24, 26, 102, 106, 110, 144, 161
Mouffe, C., 148, 152, 155

N
Negotiation, 32, 35–36, 86, 89, 113, 116–117, 149
Netherlands, the, 6, 8, 13–14, 46, 97–98, 105, 109–112, 114, 117
Animal Biotechnology Decree, 98, 101–102
legislative culture, 111, 115
poldermodel, 97, 111
political culture, 113, 117, 118
Nonet, P., 29
Norm development
ongoing processes of norm development, 6, 12, 27–28, 30, 33–35, 37–38, 95, 102, 118, 122–123, 134–135, 149, 162–163, 166
Norms
aspirational norms, 30, 39, 122–123, 134
legal norms, 10–11
open norms, 10, 12, 15, 73, 95, 110, 122, 135, 149, 162

O
Olsen, J.P. *See* March

P
Participation, public, 13–14, 32, 46, 49, 56, 62–63, 66, 75, 89–90, 106, 114, 126, 131, 149
Paula, L., 102–106, 108
People's enlightenment, principle of, 63
Pluralism, 9, 11–13, 15, 75, 90, 143–144, 149–150, 155, 163
Poldermodel, 97, 111
Policy
'no, unless' policy, 98–100, 102, 104, 110, 128
Politics of accommodation, 97, 111–113
Popular initiative, 75, 90–91
Porsborg Nielsen, A., 56, 59, 64
Positivism, legal, 25, 94–95, 116, 133, 135
Practice of legality, 19, 31, 34–35, 39
Pragmatic view of law, 6, 68–69, 71, 73, 117, 133, 146
Pre-parliamentary procedure, 77–78, 81–86, 89, 120, 122, 128, 131
Precautionary measures, 80

Principle(s)
legal principles, 18–19, 23, 30, 116-117, 127, 133
principle of people's enlightenment, 63
principle of proportionality, 52–54, 57–60, 68, 124, 127–128
principle of reciprocity, 23, 30–32, 3 4, 154
principle of shared commitments, 23, 24
Proportionality, principle of, 52–54, 57–60, 68, 124, 127–128

Q

Quality of law, 6, 15, 18–25, 31, 39, 127

R

Rawls, J., 148, 152–154
Realism, Scandinavian, 71–72, 133
Reciprocity, principle of, 23, 30–32, 34, 154
Referenda, 61, 75, 81–82, 86, 88–92, 114, 128, 132
Regulation
Danish Animal Cloning Act, 49, 51, 53, 56–57, 59
Dutch Animal Biotechnology Decree, 98, 101–102
Swiss Gene Technology Law, 75–76, 79, 81–83, 85, 120
Reich, J., 86–87, 90–92, 94
Rendtorff, J. *See* Kemp
Representation, 56, 61–62, 109, 114, 131
Rescher, N., 143, 166
Responsiveness, 21, 29–30, 109, 117, 152, 160
Right to information, 64, 78, 82, 110
Rippe, K.P., 84, 93
Rosenthal, I., 155–156, 158
Ross, A., 71, 73

S

Sandøe, P., 56, 58–59, 64, 68–69
Scandinavian realism, 71–72, 133
Schaap, A., 155
Schaber, P., 84
Scholten, P., 115
Selznick, P., 27, 29–30
Shared commitment, principle of, 23–24, 30
Shared understandings, 34, 38–39
Somsen, J., 165
Sunstein, C., 150
Swiss Animal Ethics Committee, 126–127
Swiss Ethics Committee on Non-Human Gene Technology, 77–78, 80–81, 83–85, 92–93, 95, 134–135
Switzerland, 5–6, 8, 13–14, 46
Gen-Lex motion, 76–77, 82, 91
Gene Technology Law, 75–76, 79, 81–83, 85, 120
legislative culture, 87, 93
political culture, 87

T

Tamm, D., 58–59, 70
Teubner, G., 160
Thompson, D. *See* Gutmann
Toope, S.J. *See* Brunnee
Transparency of decision-making, 56, 60, 64–65
Two-track approach, 151–152, 158–159, 162, 165–166
legal track, 162
moral track, 164

U

Undecidability, 155–157, 163

V

Validity, 14, 24, 109, 127, 133
Van Beers, B.C., 153, 157, 165
Van der Burg, W., 4–5, 11–12, 27–28, 31–34, 37–39, 158, 161, 163
Van Klink, B.M. J., 109, 116–117
Vatter, A., 88, 90
Vorstenbosch, J. *See* Ippel

W

Waldron, J., 23, 151–154, 156
Witteveen, W., 29, 31, 39, 116